Synergistic Interaction of Big Data with Cloud Computing for Industry 4.0

The idea behind this book is to simplify the journey of aspiring readers and researchers to understand the convergence of Big Data with the Cloud. This book presents the latest information on the adaptation and implementation of Big Data technologies in various cloud domains and Industry 4.0.

Synergistic Interaction of Big Data with Cloud Computing for Industry 4.0 discusses how to develop adaptive, robust, scalable, and reliable applications that can be used in solutions for day-to-day problems. It focuses on the two frontiers — Big Data and Cloud Computing – and reviews the advantages and consequences of utilizing Cloud Computing to tackle Big Data issues within the manufacturing and production sector as part of Industry 4.0. The book unites some of the top Big Data experts throughout the world who contribute their knowledge and expertise on the different aspects, approaches, and concepts related to new technologies and novel findings. Based on the latest technologies, the book offers case studies and covers the major challenges, issues, and advances in Big Data and Cloud Computing for Industry 4.0.

By exploring the basic and high-level concepts, this book serves as a guide for those in the industry, while also helping beginners and more advanced learners understand both basic and more complex aspects of the synergy between Big Data and Cloud Computing.

Innovations in Big Data and Machine Learning
Series Editors: Rashmi Agrawal and Neha Gupta

This series includes reference books and handbooks that will provide conceptual and advanced materials that cover building and promoting the field of Big Data and Machine Learning, which includes theoretical foundations, algorithms and models, evaluation and experiments, applications and systems, case studies, and applied analytics in specific domains or on specific issues.

Artificial Intelligence and Internet of Things
Applications in Smart Healthcare
Edited by Lalit Mohan Goyal, Tanzila Saba, Amjad Rehman, and Souad Larabi

Reinventing Manufacturing and Business Processes through Artificial Intelligence
Edited by Geeta Rana, Alex Khang, Ravindra Sharma, Alok Kumar Goel, and Ashok Kumar Dubey

Convergence of Blockchain, AI, and IoT
Concepts and Challenges
Edited by R. Indrakumari, R.Lakshmana Kumar, Balamurugan Balusamy, and Vijanth Sagayan Asirvadam

Exploratory Data Analytics for Healthcare
Edited by R. Lakshmana Kumar, R. Indrakumari, B. Balamurugan, Achyut Shankar

Information Security Handbook
Edited by Abhishek Kumar, Anavatti G. Sreenatha, Ashutosh Kumar Dubey, and Pramod Singh Rathore

Natural Language Processing In Healthcare
A Special Focus on Low Resource Languages
Edited by Ondrej Bojar, Satya Ranjan Dash, Shantipriya Parida, Esaú Villatoro Tello, Biswaranjan Acharya

Synergistic Interaction of Big Data with Cloud Computing for Industry 4.0
Edited by Sheetal S. Zalte-Gaikwad, Indranath Chatterjee, and Rajanish K. Kamat

For more information on this series, please visit: www.routledge.com/Innovations-in-Big-Data-and-Machine-Learning/book-series/CRCIBDML

Synergistic Interaction of Big Data with Cloud Computing for Industry 4.0

Edited by
Sheetal S. Zalte-Gaikwad, Indranath Chatterjee,
and Rajanish K. Kamat

CRC Press
Taylor & Francis Group
Boca Raton London New York

CRC Press is an imprint of the
Taylor & Francis Group, an **informa** business

MATLAB® is a trademark of The MathWorks, Inc. and is used with permission. The MathWorks does not warrant the accuracy of the text or exercises in this book. This 'book's use or discussion of MATLAB® software or related products does not constitute endorsement or sponsorship by The MathWorks of a particular pedagogical approach or particular use of the MATLAB® software.

First edition published 2023
by CRC Press
6000 Broken Sound Parkway NW, Suite 300, Boca Raton, FL 33487-2742

and by CRC Press
4 Park Square, Milton Park, Abingdon, Oxon, OX14 4RN

CRC Press is an imprint of Taylor & Francis Group, LLC

ISBN: 978-1-032-24508-9 (hbk)
ISBN: 978-1-032-24509-6 (pbk)
ISBN: 978-1-003-27904-4 (ebk)

DOI: 10.1201/9781003279044

Typeset in Times
by Newgen Publishing UK

Contents

Preface

The computing and data sector has seen tremendous changes in platform scale and application reach during the last decade. Computers, smartphones, clouds, and social media platforms need outstanding performance and a high level of machine intelligence. We are approaching a period of big data analysis and cognitive computing. The extensive use of mobile phones, storage, and computing clouds, the resurrection of artificial intelligence in practice, extended supercomputer applications, and broad deployment of Internet of Things (IoT) and Blockchain technologies are indicators of this popular trend. This book, entitled *Synergistic Interaction of Big Data with Cloud Computing*, aims to provide state-of-the-art research in the significant areas of computer science, such as Big Data, Cloud Computing, Blockchain, Machine Learning, and IoT technologies.

Big data, low-cost commodity technology, and new information management and analytic tools have created a watershed moment in data analysis history. Due to the apparent confluence of these developments, we now have the capabilities to analyze massive data sets quickly and inexpensively for the first time in history. These advancements are not hypothetical or straightforward. They are a significant step forward and have clear potential to achieve massive benefits in efficiency, productivity, revenue, and profitability.

Both big data and cloud computing play a significant part in our digital world. It is undeniable. When the two are combined, people with excellent ideas but few resources have a chance to succeed in business. These technologies also enable existing firms to use data that they already collect but could not previously analyze. Artificial intelligence and other current components of cloud infrastructure's traditional "Software as a Service" architecture enable organizations to gain insights from their big data. Businesses can take advantage of all of this for a bit of cost with a well-designed system, leaving competitors who refuse to embrace these new technologies in the dust.

The technologies that will be covered in this book are fundamental in the era of Industry 4.0. This knowledge is highly essential for readers, starting from beginner students to advanced level academicians. The confluence of big data, cloud computing, and machine learning were always examples of highly influential research. However, this book introduces the confluence of big data and blockchain technologies, which is novel in the field. We believe this will pave a path to a new way of thinking.

The Editors:
Sheetal S. Zalte-Gaikwad
Indranath Chatterjee
Rajanish K. Kamat

Editors' Biographies

Sheetal S. Zalte-Gaikwad is an assistant professor in Computer Science Department at Shivaji University, Kolhapur, India. She pursued a Bachelor of Computer Science from Pune University, India, in 2002 and a Master of Computer Science from Pune, India, in the year 2004. She earned her Ph.D. in Mobile Adhoc Network at Shivaji University. She has 14 years of teaching experience in computer science. She has published 20+ research papers in reputed international journals and conferences, including IEEE, also available online. She has also authored book chapters with Springer. Her research areas are MANET, VANET, and Blockchain Security.

Indranath Chatterjee is working as a professor in the Department of Computer Engineering at Tongmyong University, Busan, South Korea. He received his Ph.D. in computational neuroscience from the Department of Computer Science, University of Delhi, India. His research areas include computational neuroscience, schizophrenia, medical imaging, fMRI, and machine learning. He has authored and edited eight books on computer science and neuroscience published by renowned international publishers. He has published numerous research papers in international journals and conferences. He is a recipient of various global awards in neuroscience. He is currently serving as a chief section editor of a few renowned international journals, is a member of the editorial board of various international journals, and he is an advisory board member in various "Open-Science" organizations worldwide. He is presently working on several projects for government and non-government organizations as PI/co-PI, related to medical imaging and machine learning for a broader societal impact, for which he is collaborating with several universities globally. He is an active professional member of the Association of Computing Machinery (ACM, USA), the Organization of Human Brain Mapping (OHBM, USA), the Federations of European Neuroscience Society (FENS, Belgium), the Association for Clinical Neurology and Mental Health (ACNM, India), the Korean Society of Brain and Neural Science (KSBNS, Korea), and the International Neuroinformatics Coordinating Facility (INCF, Sweden).

Rajanish K. Kamat is currently working as a Vice Chancellor, Dr. Homi Bhabha State University, Mumbai, the first University of its kind in the State of Maharashtra established as per the guidelines of Rashtriya Ucchatar Shiksha Abhiyan (RUSA) to offer dynamic and demand-driven academic programmes. He has to his credit 200+ publications in journals from reputed publishing houses such as IEEE, Elsevier, and Springer in addition to 16 reference books from reputed international publishers such as Springer, UK and River Publishers, Netherlands and exemplary articles on ICT for Encyclopedia published by IGI. The books exhibit impressive download history and are placed in more than 400 Libraries all over the globe. These publications have inspired many young minds to pursue their research careers in niche technological domains. Dr. Kamat has success-fully guided 19 Ph.D. students. He has mobilized research grants to the tune of Rs. 10 Cr. from different funding agencies such as UGC, DST, MHRD, and RUSA. He is a Young Scientist awardee of Department of Science and Technology, Government of India under Fast Track Scheme.

He has been working on different committees of University Grants Commission such as the Learning Outcome Curriculum Framework (LOCF), Autonomy of HEIs, Colleges with Potential for Excellence and he is also an expert evaluator of NAAC, Bengaluru. His latest initiatives are setting up the Faculty Development Centre under Ministry of Human Resource Development's PMMMNMTT scheme in Cyber Security and Data Sciences, RUSA-Industry sponsored Centre of Excellence in VLSI System Design and the coordination of two international projects EQUAMBI (Enhancing Quality Assurance Management and Benchmarking Strategies in Indian Universities) and Internationalized Master Degree Education in Nano Electronics under the European Union's Erasmus+ scheme. Dr. Kamat is also currently working as an Adjunct Professor in Computer Science for the reputed Victorian Institute of Technology, Melbourne, Australia.

Contributors

Sourav Bardhan
Techno College of Engineering,
  Agartala, India

Himani Bhatheja
Delhi Technological University,
  New Delhi, India

Diptendu Bhattacharya
National Institute of Technology,
  Agartala, India

Sagar Chakraborty
Seacom Engineering College,
  West Bengal, India

Indranath Chatterjee
Tongmyong University, Busan,
  South Korea

Buseung Cho
Korea Institute of Science and
  Technology Information, Daejeon,
  South Korea
And University of Science and
  Technology, Daejeon, South Korea

Nivedita Das
Techno India Group, India

Dipjyoti Deb
Techno College of Engineering,
  Agartala, India

Partha Pratim Deb
Techno College of Engineering,
  Agartala, India

Rashmi Deshmukh
Shivaji University, Kolhapur,
  Maharashtra, India

Suchanda Dey
Techno College of Engineering,
  Agartala, India

Rupali Dhabarde
Shivaji University, Kolhapur,
  Maharastra, India

Sheetal S. Zalte-Gaikwad
Shivaji University, Kolhapur,
  Maharashtra, India

N. Jayanthi
Delhi Technological University,
  New Delhi, India

D.V. Kodavade
DKTE Society's Textile &
  Engineering Institute, Ichalkaranji,
  Maharashtra, India

T.V. Vijay Kumar
Jawaharlal Nehru University,
  New Delhi, India

Indra Kumari
Korea Institute of Science and Technology
  Information, University of Science and
  Technology, Daejeon, South Korea

Parijata Majumdar
Techno College of Engineering,
  Agartala, India

Ruchismita Majumder
Techno College of Engineering,
  Agartala, India

Sanjoy Mitra
Tripura Institute of Technology,
  Narsingarh, India

L. Steffina Morin
SRM Institute of Science and
  Technology, Tamil Nadu, India

Debankita Mukhopadhyay
Techno College of Engineering,
  Agartala, India

P. Murali
SRM Institute of Science and
  Technology, Tamil Nadu, India

Mary Olawuyi Ojo
Asarawa State University, Nigeria
and Nigerian Postal Service, Abuja,
  Nigeria

Debasmita Pal
Techno College of Engineering,
  Agartala, India

Kisor Ray
Techno India University, India

Sourabarna Roy
Techno College of Engineering,
  Agartala, India

Jungsuk Song
Korea Institute of Science and
  Technology Information, University
  of Science and Technology, Daejeon,
  South Korea

Udeechee
Jawaharlal Nehru University
New Delhi, India

1 Big Data Based on Fuzzy Time-Series Forecasting for Stock Index Prediction

*Ruchismita Majumder
Department of Computer Science and Engineering,
Techno College of Engineering, Agartala

Partha Pratim Deb
Department of Computer Science and Engineering,
Techno College of Engineering, Agartala

Diptendu Bhattacharya
Department of Computer Science and Engineering,
National Institute of Technology, Agartala

*Corresponding author.

CONTENTS

DOI: 10.1201/9781003279044-1

1

1.1 INTRODUCTION

Forecasting Methods uses fuzzy time series to manage the uncertainties faced with TAIEX (Taiwan Stock Exchange Capitalization Weighted Stock Index) forecasting, where alternate quotes of forecasting are used (Yanpeng Zhang et al. 2020). If Stock Index Prediction is excessive, income may be increased. To ease the procedure, it is essential to expand a correct technique in the prediction of inventory guides, inflicting the prediction trouble of inventory volatility (Shen and Shafiq 2019). The forecast price at time $t + 1$ is described because the information factor at time t, plus the forecast variation (Pal et al. 2019). In this paper, the proposed technique is carried out using statistics from the TAIEX Capitalization, with the NASDAQ (National Association of Securities Dealers Automated Quotation). In earlier years, scholars were not only dedicated to stock-price-related study but also tried to examine stock market dealings such as volume burst risks in such uncertainties, which increased the stock index study research sector and indicated that this research domain still has high potential. The proposed method (Chen and Jian 2017; Chen et al. 2015) shows a process using TAIEX forecasting based on different factors like possibilities of fuzzy logics along with two-factor second-order fuzzy inclination logical relationship groups. Chen and Chen (2011) explain various prediction implementing with TAIEX using FTS with different fuzzy variation groups. The difference in amount of adjacent historical data in order to generate fuzzy variation groups for the fuzzy time series, and the technique was then applied in order to forecast the TAIEX (Abhishekh 2019). A stock market also known as equity market, or share market, is the combination of investors and shareholders of stocks (also called shares), that shows (Eapen et al. 2019) Rights claims on industries; this consists of securities operated privately, over stocks of private organizations which are traded to different stages (Bhanja and Abhishek 2019).

Brookers and sellers play the vital role in stock market investment where most of the investment is done by talented minds who know the market more prominently. Each share is categorized on the basis of the location of the organization (Yu 2005). For example, Unilever is based in the United States and traded on the SIX Swiss Exchange (Idrees et al. 2019), and this is measured as a portion in the Swiss stock market (S. Zhang 2016).

Previously, several time-series prediction models were planned and used to help make financial estimates; as such the autoregressive restricted standard, the autoregressive moving average model, and the autoregressive integrated moving average model (Ince and Trafalis 2008).

In 1875, the stock market came into effect in India the moment the Bombay stock market was recognized as a traders association. The stock market is a good example of any neighborhood market where items are bought and sold. Like any marketplace, goods are collected and traded. The stock market helps in determining the stock during the day along with auction prices. Buyers compete with each other and sell the stock at a price. Similarly, investors participate with one another, biding the best price for the day or the lowest price to sell the share.

The innovation of this planned model is to feature engineering sideways with a different system in place of any other model. After the success of the feature extension method, it helps in collaborating with different feature exclusion technique, which helps in allowing machine-learning algorithms to attain accuracy marking short term price prediction (Bisht and Kumar 2016). It also showed the accuracy of the planned feature allowance as feature engineering. In recent years, scholars were more dedicated on stock price-related investigation and also worked in the analyzation of stock market connection in various ways that help in expansion of stock market.

Lee (2009) used Support Vector Machine (SVM) with different fusion feature variety techniques to carry forward the analyzation of share market. The dataset for the proposed paper substitute dataset of TAIEX Index in Taiwan Economic Journal Database (TEJD) in 2008 (Shen and Shafiq 2020). The variety of different selection supported sequential forward search (SSFS) which contributed to the role of the packaging (Fang et al. 2021). Another merit of the proposed paper is that it was intended as a thorough method of constraint alteration with presentation under different constraint standards (Eapen et al. 2019). The vibrant construction of the feature selection prototypical is also experimental to the various stages of prototypical configuring (Sirignano et al. 2019).

Stock Market Forecasting is used to forecast the upcoming worth of the economic stocks of an organization (Gangwar and Kumar 2012). The current trend in stock market forecasting machineries is the working of machine-learning techniques making forecasting depending on the ethics of existing stock market guides by working out their prior principles.

The data in view of the variability of the different foundations and the unreliablility of the statistics produced for gathering progression, where most of the gathered information comprehend partial, inaccurate, (Hsu 2013) and vague chronicles, result in making the initialization a lengthy method for machine-learning techniques. In order to diminish the prediction fault to 50% of various prevailing methods intermission-created fuzzification, CFTS (Combination of the Input Variables) substitutes the progression of fuzzification. The asset of this research work is that the author appraised equally the probabilistic distance constructed and several feature selection methods (Thakur and Kumar 2018; Shih et al. 2019). Also, the author accomplished the appraisal grounded on various datasets, which armored the asset of the research work reducing the accurateness of the paper.

The rest of the paper is designed in this manner. In Section 1.2, brief evaluations of some basic notions of fuzzy time series from (Jiang et al. 2018). In Section 1.3, the methodology of the paper is discussed. The paper comprises of three methods: namely TAIEX, BSE and KOSPI grounded on fuzzy time series. In Section 1.4, a comparison of RMSE values is checked along with the actual and forecasted data of the planned technique with the prevailing procedures. Section 1.5 contains the conclusion.

1.2   DISCUSSION ON FUZZY TIME-SERIES PREDICTION

In this section, an innovative technique to predict the TAIEX (Figures 1.2 – 1.7) created on fuzzy time series and fuzzy distinction groups, where the main method comprises of TAIEX where other methods are also checked.

The study shows the compensations of FTS applied to resolve the linguistic term problem. The FTS notions are discussed below. A fuzzy set A is well-defined in the universe of discourse $U = \{u_1, u_2, \ldots, u_m\}$ as

$$A = fA(u_1)/u_1 + fA(u_2)/u_2 + \cdots + fA(u_m)/u_m, \tag{1}$$

Where, fA = connection meaning of the fuzzy set A_i, $fA_i(u_i)$ = shows the degree of connection of u_i appropriate in the fuzzy set A_i $fAi(u_i) \in [0, 1]$ and $1 \leq j \leq$ m.

As the fuzzy relationship $R(S-1, s)$ exists, such that $F(s) = F(s-1) \odot R(s-1, s)$ where the sign "$\odot$" indicates the max–min configuration operator, then $F(s-1) \rightarrow F(s)$ denoted by the fuzzy logical relationship. Let $F(s-1) = A_i$ and let $F(s) = A_j$. The connection between $F(s-1) \rightarrow F(s)$ can be indicated by the fuzzy logical relationship "$Ai \rightarrow Aj$," where Ai and Aj are the LHS and RHS respectively of the fuzzy logical relationship "$Ai \rightarrow Aj$," respectively.

Grouping fuzzy logical relationships have the similar LHS into a FLRG. For example, adopt the following fuzzy logical relationships exist:

$$Ai \rightarrow Aj\ a\ Ai \rightarrow Aj\ b \ldots Ai \rightarrow Ajm.$$

Which helps in fuzzy logical relationships grouping into FLRG.

The trial outcomes illustrate the proposed techniques on fuzzy time-series model with various alternatives accomplishes various models where reliant on the additional methods followed in the paper.

1.3 METHODOLOGY

The procedure implied in this research using big data arranged for fuzzy time-series prediction for stock index forecasting which consists of stock market prediction evolution of the last six years using big data.

The paper consists of three different kinds of data indices for our experimental purpose. Where three data TAIEX, KOSPI, BSE are used for MFTS (Membership Functions). Six years of data is used for the training phase from 2015 to 2020 from where 30% of data is used for the testing phase irrespective of any methods adopted.

Furthermore, the appraised the accurateness of the calculations via various training set and a test set. The verdicts from the literature review are concise as follows:

1) The appropriately biased tactic to the FLRs of fuzzy time series helps in correctly imitating the need of different distinct FLR resulting in toughening the forecasting accurateness.
2) The exact involvement notch of the type-1 fuzzy sets indicates how it is unable to deal with glitches involving vagueness, such as oscillations of stock catalogues.

Type-2 fuzzy sets, on the further way permits correlated membership degrees to be inexact, that are able to grip such problems.

The Bombay Stock Exchange (BSE commonly recognized as the BSE), is an Indian stock exchange which is situated on Dalal Street in Mumbai (Chen et al. 2015). It was recognized in the year 1875 by cotton merchant Premchand Roychand, a Rajasthani Jain manufacturer. The BSE is the oldest stock exchange in Asia, which is also the tenth oldest in the world. The datasets from 2015 to 2020 are being examined in this paper (Dhruv Devani 2022).

In accordance with earlier works, different scholars consider both financial domain information along with the procedural approaches on stock data which helped using instructions to strain the high-quality shares (Chen and Jian, 2017). Though in order to certify the best presentation of the prediction paper, let's just look into the statistics primarily (Abhishekh and Kumar, 2020). The paper comprises of a large number of structures in the underdone data; if the involvement of all the structures into the deliberation, it will not only harshly surge the computational intricacy but will also cause side effects if unsupervised learning is being presented

Stage 1: U, $U = \left[R_{min} - R_1, R_{max} + R_2 \right]$, R_{min} R_{max}

Indicates the minimum and maximum values;

R_1 and R_2 are two positive actual values to which partitions the universe of discourse U into m intervals $U_1, U_2, \ldots,$ and U_m of equal length.

Stage 2: Define the linguistic terms $A_1, A_2, \ldots, A_m$ represented by fuzzy sets of the main factor, shown as follows

$$A_1 = 1/u_1 + 0.5/u_2 + 0/u_3 + \cdots + 0/u_m - 2 + 0/u_m - 1 + 0/u_m \qquad (2)$$

$$A_2 = 0.5/u_1 + 1/u_2 + 0.5/u_3 + \cdots + 0/u_m - 2 + 0/u_m - 1 + 0/u_m \qquad (3)$$

$$A_n - 1 = 0/u_1 + 0/u_2 + 0/u_3 + \cdots + 0.5/u_m - 2 + 1/u_m - 1 + 0.5/u_m \qquad (4)$$

$$A_n = 0/u_1 + 0/u_2 + 0/u_3 + \cdots + 0/u_m - 2 + 0.5/u_m - 1 + 1/u_m \qquad (5)$$

where $U_1, U_2, \ldots,$ and U_m are intermissions gained in Stage 1.

Stage 3: Fuzzification of individually training data of the main influence into a fuzzy set defined in Stage 2. The training data of the particular factor of training day belongs to U_i and the extreme value of the fuzzy set A_i happening at interval U_i where $1 \leq i \leq n$, where predicted data is fuzzified into A_i.

Stage 4: The fuzzified training data of the key factor is gained to paradigm fuzzy relationship in Stage 3.

Stage 5: Fuzzification of the disparities among the together training data of the main factor, respectively, and finally grouping the fuzzy logical relationships of the main factor obtained in Stage 4.

Stage 6: It is the difference in the fuzzification of the secondary factor of the training data and examining the fuzzy differences appeared in the fuzzy variation groups obtained in Stage 5.

Stage 7: For exact prediction, each fuzzy variation group contains dissimilar datasets to accomplish the calculation.

In the proposed paper, to equivalent the experimental consequences of the proposed method along with the one of the procedures offered in (Chen 1996) (Huarng et al. (2007) (Hui-Kuang Yu et al. 2008) and Huarng et al. 2007), adopting six years of data and splitting them for training and testing, which is as similar as the untried environment of Chen (1996), Huarng et al. (2007), Yu and Huarng (2008) and Huarng et al., (2008) for several years, the training dataset was from 2015 to 2020 and the rest are analyzed as the testing dataset.

Step 1: From (K. K. Gupta et al., 2018), here the minimum result and the maximum result of the training data of the TAIEX of the year 2015 is 6316.89 and 7134.02 respectively. Henceforth, $D_1 = 26.87$, and let $D_2 = 67.4$.

So, the universe in discourse $u = [6900, 7400]$. In lieu of comparison of the experimental results of the proposed method with the one obtainable in Nurmaini and (2017), the span of each intermission is same as the one shown in Nurmaini Chusniah (2017). Therefore, assuming the length of each intermission in the universe in discourse u be 200. So, the universe in discourse U can be divided into 16 intervals u_1, u_2 ...., and u_{16}, discussed below: $u_i = \left[6900 + (i - 1) \times 200, 6900 + i \times 200\right)$

Step 2: Depending on the intervals obtained in Step 1, the fuzzy set A_i ($i = 1, 2, ...,$ 16) as follows:

$$A_1 = 1/u_1 + 0.5/u_2 + 0/u_3 + \cdots + 0/u_{14} + 0/u_{15} + 0/u_{16} \qquad (6)$$

$$A_2 = 0.5/u1 + 1/u_2 + 0.5/u_3 + 0/u_4 + 0/u_5 + 0/u_6 + 0/u_7; \qquad (7)$$

$$A_{16} = 0/u_1 + 0/u_2 + 0/u_3 + \cdots + 0/u_{14} + 0.5/u_{15} + 1/u_{16} \qquad (8)$$

Step 3: TAIEX on the year 2015 is 6341.52. Depending on the various fuzzy sets discussed in Step 2, the value is "6341.52" is fuzzified into the fuzzy set A_8.

Step 4: Constructed on the values of Step 3, as the fuzzified TAIEX of training year 2017 is A_8 and as the fuzzified TAIEX of 2019 is A_9.

1.4 RESULTS AND DISCUSSION

The discussion on this section summarizes the results of various equations and graphs between actual and forecasted data. Three methods are followed in order to validate the prediction accurateness of the proposed technique. An evaluation is drawn and shown with various additional predicting techniques, which utilizes the root-mean-square error (RMSE) to portion prediction accurateness revealed in Table 1.1.

$$\sqrt{\frac{\sum_{i=1}^{n}(actual\ value_i - forecasted\ value_i)^2}{n}} \tag{9}$$

Where denoting n as the number of predicted values. The RMSE created by various techniques which are utilized to analyse the typical RMSE comprising the standard deviation (SD), which is being imperilled to the numerical investigation where in the constant of variation (CV) is planned. Earlier techniques only measured as the regular RMSE while conclusive of which technique will help result in more accurate predicting presentation. However, the changes may make the process more problematic to prove if the predicting technique is significantly improved than any other techniques, which is very similar to that of the average which may have dissimilar SD. Henceforth, this is more significant towards the employment of numerical investigation in order to confirm predicting presentation. Diverse numerical trial is selected as the numerical investigation technique for examining alterations in the predicting presentation among the planned technique and additional prevailing procedures. If the numerical important value is initiated around smaller RMSE values, then the predicting presentation of the planned technique is restored than that of other techniques.

1.4.1 TAIEX FORECASTING

The Taiwan Stock Exchange Corporation (TWSE) was established in 1961 and started operating from 9 February 1962.The current chairman of TWSE is Hsu Jan-yau and the president is Lee Chi-Hsien. Taiwan Capitalization Weighted Stock Index (TAIEX) measures the performance of stocks listed in TWSE.

TABLE 1.1
RMSE Values Comparison Chart and Average Values for Different Methods

| Method | Year | | | | | | |
	2015	2016	2017	2018	2019	2020	Average
TAIEX	278.42	232.35	346.25	337.52	363.40	432.21	347.19
BSE	838.07	992.16	1152.43	1169.54	1138.99	1381.38	1112.09
KOSPI	70.94	64.76	78.22	70.83	71.56	82.89	73.2

Graphical Representation of Actual and Forecasted Data from 2015 to 2020

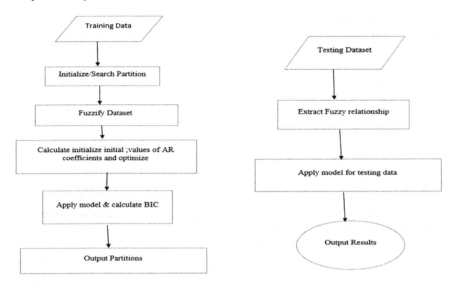

FIGURE 1.1   Fuzzy time-series prediction flowchart.

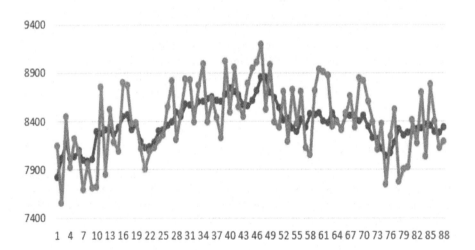

FIGURE 1.2   Actual vs. forecasted for TAIEX 2015.

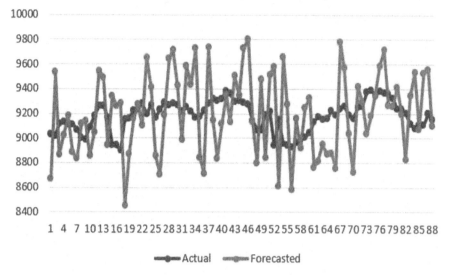

FIGURE 1.3   Actual vs. forecasted for TAIEX 2016.

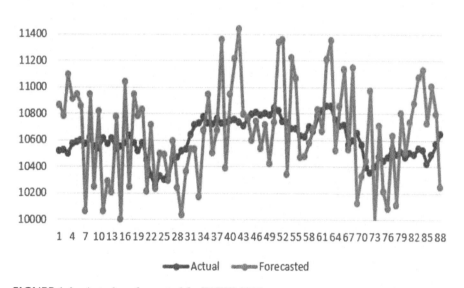

FIGURE 1.4   Actual vs. forecasted for TAIEX 2017.

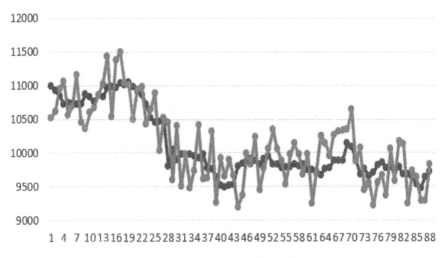

FIGURE 1.5 Actual vs. forecasted for TAIEX 2018.

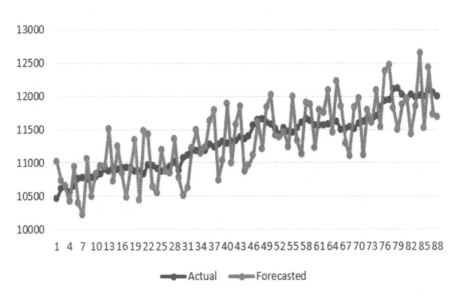

FIGURE 1.6 Actual vs. forecasted for TAIEX 2019.

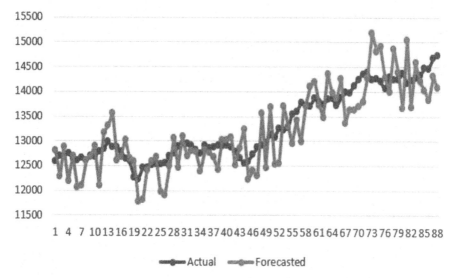

FIGURE 1.7 Actual vs. forecasted for TAIEX 2020.

TABLE 1.2
Comparison of Average RMSE Values for TAIEX Method

Methods	Year						Average RMSE
	2015	2016	2017	2018	2019	2020	
Huarng et al's	167.8	112.68	19.67	154.82	199.45	199.32	142.29
Chen's Fuzzy Time Series	169.7	175.77	192.67	157.45	179.92	197.69	178.86
Univariate Conventional Regression	153.5	259.45	213.67	200	230.19	441.72	249.75
Univariate Neural Network	78	97.7	232.56	339.60	451.67	321.7	253.53
Bivariate Conventional Regression	109.78	216.5	347.39	312.98	219.7	431.79	273.02
Bivariate Neural Network	289.32	190.8	172.98	231.87	321.7	421.7	223.14
Proposed Model	278.42	232.35	346.25	337.52	363.40	432.21	347.19

Table 1.2 shows all the comparison of RMSE values of different years along with the Average RMSE value as compared with earlier researches. (Hui-Kuang Yu et al. 2008), Chen (1996) and Yu and Huarng (2010) show the earlier research completed with method TAIEX from the year 2015 to 2020.

1.4.2 BSE FORECASTING

The BSE is a non-restricted market-weighted stock market directory of 30 different deep-rooted and monetarily self-governing companies that are registered on the Bombay Stock Exchange. The 30 main companies are some of the major and most vigorously dealt stocks, which represent many of numerous manufacturing sectors of the Indian economy. The BSE adapts its arrangement to reflect present market scenarios. The index is assumed depending on a non-restricted capitalization technique, a distinction of the market capitalization method.

Graphical Representation of Actual and Forecasted Data from 2015 to 2020

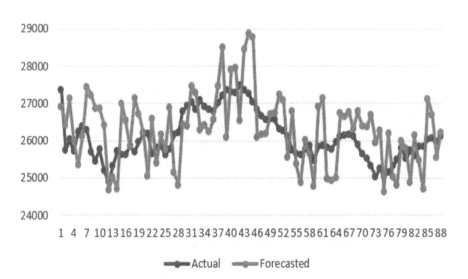

FIGURE 1.8   Actual vs. forecasted for BSE 2015.

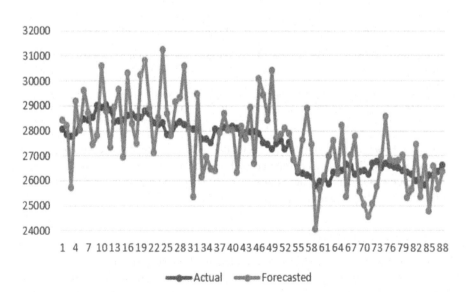

FIGURE 1.9   Actual vs. forecasted for BSE 2016.

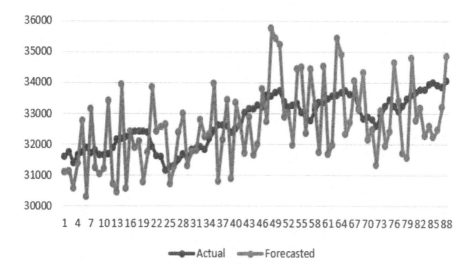

FIGURE 1.10   Actual vs. forecasted for BSE 2017.

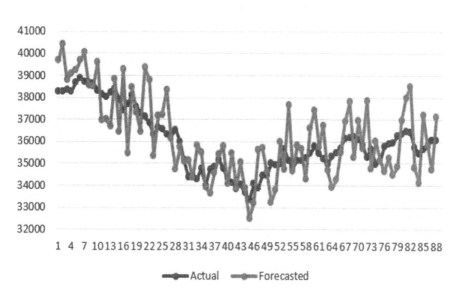

FIGURE 1.11   Actual vs. forecasted for BSE 2018.

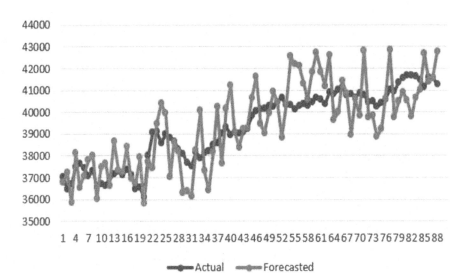

FIGURE 1.12   Actual vs. forecasted for BSE 2019.

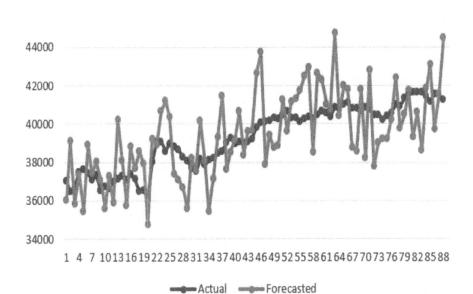

FIGURE 1.13   Actual vs. forecasted for BSE 2020.

TABLE 1.3
Comparison of Average RMSE Values for BSE Method

Methods	Year						Average RMSE
	2015	2016	2017	2018	2019	2020	
Huarng et al's	782.9	894.76	675.89	789.8	828.9	987.43	826.61
Chen's Fuzzy Time Series	672.76	782.7	829.65	1179.78	965.9	1152.21	930.5
Univariate Conventional Regression	544.29	1167.73	577.7	453.7	765.98	875.4	730.8
Univariate Neural Network	839.06	987.41	1234.9	873.9	356.99	241.97	755.70
Bivariate Conventional Regression	431.78	567.78	356.89	563.92	1357.54	543.74	636.94
Bivariate Neural Network	218.7	980.98	548.7	678.54	1129.89	342.9	649.95
Proposed Model	838.07	992.16	1152.43	1169.54	1138.99	1381.38	1112.09

Table 1.3 shows the comparison of various studies from the year 2015 to 2020, along with the Average RMSE value with method BSE.

1.4.3 KOSPI FORECASTING

The Korea Composite Stock Price Index commonly known as KOSPI deals with common shares of Stock Market Division—earlier, Korea Stock Exchange—of the Korea Exchange. This works as representative stock market index of South Korea.

KOSPI was announced in 1983 with the base value of 100 as of 4 January 1980. It is designed based on market capitalization.

Graphical Representation of Actual and Forecasted Data from 2015 to 2020

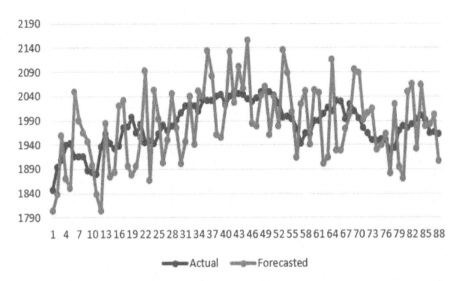

FIGURE 1.14   Actual vs. forecasted for KOSPI 2015.

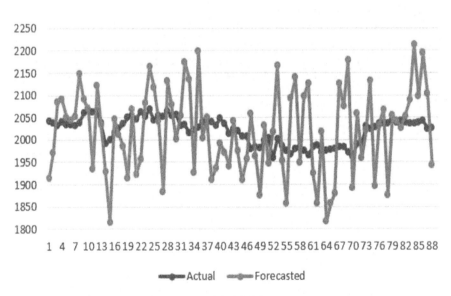

FIGURE 1.15   Actual vs. forecasted for KOSPI 2015.

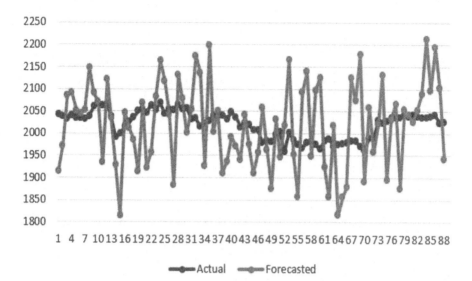

FIGURE 1.16   Actual vs. forecasted for KOSPI 2017.

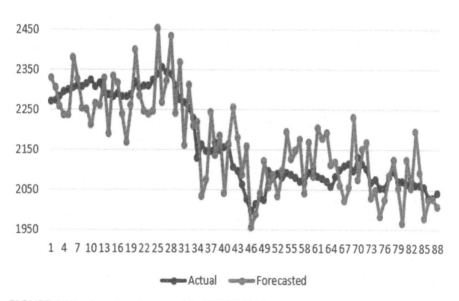

FIGURE 1.17   Actual vs. forecasted for KOSPI 2018.

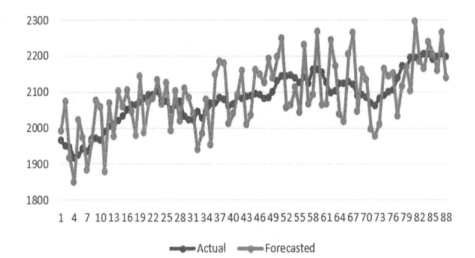

FIGURE 1.18   Actual vs. forecasted for KOSPI 2019.

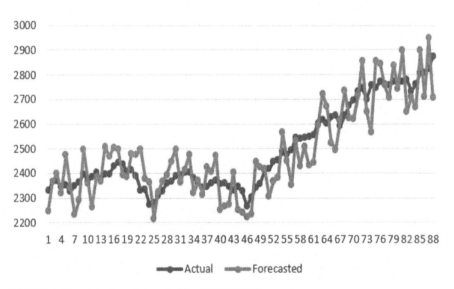

FIGURE 1.19   Actual vs. forecasted for KOSPI 2020.

TABLE 1.4

Comparison of Average RMSE Values for KOSPI Method

	Year						
Methods	2015	2016	2017	2018	2019	2020	Average RMSE
Huarng et al's	62.9	18.76	57.89	49.8	88.9	98.43	62.78
Chen's Fuzzy Time Series	67.76	32.7	29.65	19.78	65.9	52.21	44.6
Univariate Conventional Regression	44.29	17.73	87.7	45.7	76.98	75.4	57.9
Univariate Neural Network	88.06	47.41	34.9	73.9	36.99	41.97	53.03
Bivariate Conventional Regression	31.78	57.78	56.89	56.92	57.54	54.74	52.60
Bivariate Neural Network	21.7	80.98	48.7	67.54	69.89	42.9	55.28
Proposed Model	70.94	64.76	78.22	70.83	71.56	82.89	73.2

Table 1.4 shows all the comparison of RMSE values of different years along with the Average RMSE value as compared with earlier researches. Huarng and Yu (2007) Chen (1996), and Yu and Huarng (2010) show the earlier studies being followed with method KOSPI from the year 2015 to 2020.

The result shows the fluctuations in various time series of the paper. All the equations are done for the purpose of the prediction saying which is more accurate and error free. Fuzzification of the disparities among various training datasets of the key issues, which correspondingly, groups the fuzzy logical relationships of the key factor depending on the comparisons.

Figure 1.1 shows the flowchart showing the various functions done in the prediction of the actual and forecasted data. In the flowchart the whole dataset is created for training and testing where the process is started from an initial position. A fuzzified dataset is created which calculates the initial values of different sorts which helps to apply the model and calculate the BIC.

This paper executes as a vigorous fuzzy time-series prediction model depending upon time-series gathering, assembling three main contributions. In order to validate the presentation of the planned prototypical, trials are made on the dataset for TAIEX, BSE and KOSPI for every particular year, where data from 2015 to 2020 are used to paradigm the training set and the data as the testing set.

An evaluation of RMSE values is shown in Table 1.1 where the value of each year fluctuates also showing an average value for every year starting from 2015 to 2020.

A graphical representation between the actual vs. forecasted data is represented in the Table 1.1 and the figures illustrate the data as follows:

Figure 1.2 demonstrates the actual vs. forecasted data for the year 2015 using method TAIEX.

Figure 1.3 represents the actual vs. forecasted data for the year 2016 using method TAIEX.

Figure 1.4 illustrates the actual vs. forecasted data for the year 2017 using method TAIEX.

Figure 1.5 shows the actual vs. forecasted data for the year 2018 using method TAIEX.

Figure 1.6 displays the actual vs. forecasted data for the year 2019 using method TAIEX.

Figure 1.7 illustrates the actual vs. forecasted data for the year 2020 using method TAIEX.

Figure 1.8 demonstrates the actual vs. forecasted data for the year 2015 using method BSE.

Figure 1.9 represents the actual vs forecasted data for the year 2016 using method BSE.

Figure 1.10 illustrates the actual vs forecasted data for the year 2017 using method BSE.

Figure 1.11 shows the actual vs forecasted data for the year 2018 using method BSE.

Figure 1.12 displays the actual vs forecasted data for the year 2019 using method BSE.

Figure 1.13 illustrates the actual vs forecasted data for the year 2020 using method BSE.

Figure 1.14 demonstrates the actual vs forecasted data for the year 2015 using method KOSPI.

Figure 1.15 represents the actual vs forecasted data for the year 2016 using method KOSPI.

Figure 1.16 illustrates the actual vs forecasted data for the year 2017 using method KOSPI.

Figure 1.17 shows the actual vs forecasted data for the year 2018 using method KOSPI.

Figure 1.18 displays the actual vs forecasted data for the year 2019 using method KOSPI.

Figure 1.19 illustrates the actual vs forecasted data for the year 2020 using method KOSPI.

The final RMSE values for **TAIEX** from 2015 to 2020 are 278.42, 232.35, 346.25, 337.52, 363.40, 432.21 respectively. The final RMSE values for **BSE** from 2015 to 2020 is 838.07, 992.16, 1152.43, 1169.54, 1138.99, 1381.38 respectively. The final RMSE values for **KOSPI** from 2015 to 2020 is 70.94, 64.76, 78.22, 70.83, 71.56 and 82.89 respectively.

1.5 CONCLUSION

The purpose of the paper is to plan a novel fuzzy time-series model grounded on coarse set rule introduction for prediction on stock index, in which the proposed technique applied various calculation processes generated to forecast rules, which makes it dissimilar from earlier fuzzy time-series model using big data and hired adaptive expectation technique to reinforce forecasting presentation. The outcomes have revealed the planned model with better and different models that predicts the presentation in terms of correctness and accuracy. Unlike earlier research, the proposed study customs many numerical techniques to inspect the predicting performance of both the planned technique and other approaches. The first case of TAIEX forecasting, which was the proposed method and was developed to suggest better forecasting presentation than the other methods offered before 2017 (Kai et al. 2015), and shows no difference after 2017. Also, no statistically important alteration is found amid the methods established after 2017. One probable purpose of this paper is that the proposed method only practices one factor to predict the TAIEX, while other methods use one or more factors, which helps in recovering the predicting presentation. However, the proposed method only uses one factor, an outdated method for partition of the fuzzy methods, and second-order historical data to predict the stock index and from predicting the consequences of this study using three databases, TAIEX, BSE, and KOSPI, the proposed method is initiated to suggest better predicting performance or not statistically dissimilar than other techniques. Also, for replacement of shapes an original high order time-series gathering algorithm is evaluated, which paradigms a more appropriate linear model than other cluster-based models. The planned fuzzy methods add FTS-based model on the TAIEX, BSE and are equated with three different methods and less error which is more accurate. The model is related with other FTS-based model on the TAIEX, BSE and KOSPI which makes the paper precise and gives better forecasting accuracy compared to earlier reports. The forecasting model based on FTS shows that the planned forecasting model can work with partial and inexact data yet providing with the accurate result. This paper tends to work accurately in the near future reducing the error more effectively and more accurately.

REFERENCES

Abhishekh, S and Kumar, Sanjay (2020). Handling higher order time series forecasting approach in intuitionistic fuzzy environment. *Journal of Control and Decision*, 7(4), 1–18. doi:10.1080/23307706.2019.1591310

Bisht, K. and S. Kumar (2016) Fuzzy time series forecasting method based on hesitant fuzzy sets Expert Systems with Applications DOI: 10.1016/j.eswa.2016.07.044bi

Chen, Kai; Zhou, Yi; Dai, Fangyan (2015). [IEEE 2015 IEEE International Conference on Big Data (Big Data) – Santa Clara, CA, USA (2015.10.29-2015.11.1)] 2015 IEEE International Conference on Big Data (Big Data) – A LSTM-based method for stock returns prediction: A case study of China stock market, 2823–2824. doi:10.1109/BigData.2015.7364089

Chen, Shyi-Ming (1996). Forecasting enrolments based on fuzzy time series, 81(3), 311–319. doi:10.1016/0165-0114(95)00220-0

Chen, Shyi-Ming; Chen, Chao-Dian (2011). TAIEX Forecasting Based on Fuzzy Time Series and Fuzzy Variation Groups. IEEE Transactions on Fuzzy Systems, 19(1), 1–12. doi:10.1109/TFUZZ.2010.2073712

Chen, Shyi-Ming; Jian, Wen-Shan (2017). Fuzzy forecasting based on two-factors second-order fuzzy-trend logical relationship groups, similarity measures and PSO techniques. *Information Sciences*, 391–392, 65–79. S0020025516316036. doi:10.1016/j.ins.2016.11.004

Dhruv Devani, Margin Patel (2022) Stock Market (BSE) Prediction Using Unsupervised Sentiment Analysis and LSTM: A Hybrid Approach DOI: 10.1007/978-981-16-6407-6_8

Eapen, Jithin; Bein, Doina; Verma, Abhishek (2019). [IEEE 2019 IEEE 9th Annual Computing and Communication Workshop and Conference (CCWC) – Las Vegas, NV, USA (2019.1.7-2019.1.9)] 2019 IEEE 9th Annual Computing and Communication Workshop and Conference (CCWC) – Novel Deep Learning Model with CNN and Bi-Directional LSTM for Improved Stock Market Index Prediction, (), 0264–0270. doi:10.1109/CCWC.2019.8666592

Eapen, Jithin; Bein, Doina; Verma, Abhishek (2019). [IEEE 2019 IEEE 9th Annual Computing and Communication Workshop and Conference (CCWC) – Las Vegas, NV, USA (2019.1.7-2019.1.9)] 2019 IEEE 9th Annual Computing and Communication Workshop and Conference (CCWC) – Novel Deep Learning Model with CNN and Bi-Directional LSTM for Improved Stock Market Index Prediction, 0264–0270. doi:10.1109/CCWC.2019.8666592

Fang, W., Xue, Q., Shen, L., & Sheng, V. S. (2021). Survey on the Application of Deep Learning in Extreme Weather Prediction. *Atmosphere*, 12(6), 661. doi:10.3390/atmos12060661

Gangwar Singh, Sukhdev; Kumar, Sanjay (2012). Partitions based computational method for high-order fuzzy time series forecasting, 39(15). doi: 10.1016/j.eswa.2012.04.039

Gupta, Krishna Kumar; Kumar, Sanjay (2018). Hesitant probabilistic fuzzy set-based time series forecasting method. *Granular Computing*. doi:10.1007/s41066-018-0126-1

Hsu, Chih-Ming (2013). A hybrid procedure with feature selection for resolving stock/futures price forecasting problems, *Neural Computing and Applications* 22(3–4), 651–671. doi:10.1007/s00521-011-0721-4

Huarng, Kun-Huang; Hui-Kuang Yu, Tiffany; Wei Hsu, Yu (2007). A Multivariate Heuristic Model for Fuzzy Time-Series Forecasting, *IEEE Trans Syst Man Cybern B Cybern* 37(4), 0–846. doi:10.1109/tsmcb.2006.890303

Hui-Kuang Yu, Tiffany; Huarng, Kun-Huang (2008). A bivariate fuzzy time series model to forecast the TAIEX, *Expert Systems with Applications* 34(4), 2945–2952. doi: 10.1016/j.eswa.2007.05.016

Hui-Kuang Yu, Tiffany; Huarng, Kun-Huang (2010). Corrigendum to "A bivariate fuzzy time series model to forecast the TAIEX" *Expert Systems with Applications* 34 (4) (2010) 2945–2952], 37(7), 5529–0. doi:10.1016/j.eswa.2010.03.063

Idrees, Sheikh Mohammad; Alam, M. Afshar; Agarwal, Parul (2019). A Prediction Approach for Stock Market Volatility Based on Time Series Data. *IEEE Access*, 1–1. doi:10.1109/ACCESS.2019.2895252

Ince, Huseyin; Trafalis, Theodore B. (2008). Short term forecasting with support vector machines and application to stock price prediction. *International Journal of General Systems*, 37(6), 677–687. doi:10.1080/03081070601068595

Jiang, Joe-Air; Syue, Chih-Hao; Wang, Chien-Hao; Wang, Jen-Cheng; Shieh, Jiann-Shing (2018). An Interval Type-2 Fuzzy Logic System for Stock Index Forecasting Based on Fuzzy Time Series and a Fuzzy Logical Relationship Map. *IEEE Access*, 1–1. doi:10.1109/ACCESS.2018.2879962

Lee, Ming-Chi (2009). Using support vector machine with a hybrid feature selection method to the stock trend prediction, *Expert Systems with Applications* 36(8), 10896–10904. doi: 10.1016/j.eswa.2009.02.038

Nurmaini, Siti; Chusniah, (2017). [IEEE 2017 International Conference on Electrical Engineering and Computer Science (ICECOS) – Palembang (2017.8.22-2017.8.23)] 2017 International Conference on Electrical Engineering and Computer Science (ICECOS) – Differential drive mobile robot control using variable fuzzy universe of discourse, 50–55. doi:10.1109/ICECOS.2017.8167165

Pal, Shanoli Samui; Kar, Samarjit (2019). Time series forecasting for stock market prediction through data discretization by fuzzistics and rule generation by rough set theory. *Mathematics and Computers in Simulation*, S0378475419300011. doi: 10.1016/j.matcom.2019.01.001

Samit, Bhanja, Abhishek, Das (2019) *Impact of Data Normalization on Deep Neural Network for Time Series Forecasting.* doi.org/10.48550/arXiv.1812.05519

Shen, Jingyi; Shafiq, M. Omair (2019). [IEEE 2019 18th IEEE International Conference on Machine Learning and Applications (ICMLA) – Boca Raton, FL, USA (2019.12.16-2019.12.19)] 2019 18th IEEE International Conference on Machine Learning and Applications (ICMLA) – Learning Mobile Application Usage – A Deep Learning Approach, 287–292. doi:10.1109/icmla.2019.00054

Shen, Jingyi; Shafiq, M. Omair (2020). Short-term stock market price trend prediction using a comprehensive deep learning system. *Journal of Big Data*, 7(1), 66. doi:10.1186/s40537-020-00333-6

Shih, Dong-Her; Hsu, Hsiang-Li; Shih, Po-Yuan (2019). [IEEE 2019 IEEE 4th International Conference on Cloud Computing and Big Data Analysis (ICCCBDA) – Chengdu, China (2019.4.12-2019.4.15)] 2019 IEEE 4th International Conference on Cloud Computing and Big Data Analysis (ICCCBDA) – A Study of Early Warning System in Volume Burst Risk Assessment of Stock with Big Data Platform, 244–248. doi:10.1109/ICCCBDA.2019.8725738

Sirignano, Justin; Cont, Rama (2019). Universal features of price formation in financial markets: perspectives from deep learning. *Quantitative Finance*, 19(9), 1–11. doi:10.1080/14697688.2019.1622295

Thakur, Manoj; Kumar, Deepak (2018). A hybrid financial trading support system using multi-category classifiers and random forest. *Applied Soft Computing*, 67, S1568494618301224–. doi:10.1016/j.asoc.2018.03.006

Yu, Hui-Kuang (2005). Weighted fuzzy time series models for TAIEX forecasting, 349(3–4), 609–624. doi: 10.1016/j.physa.2004.11.006

Zhang, S. (2016) Architectural complexity measures of recurrent neural networks, (NIPS) https://doi.org/10.48550/arXiv.1602.08210

Zhang, Yanpeng; Qu, Hua; Wang, Weipeng; Zhao, Jihong (2020). A Novel Fuzzy Time Series Forecasting Model Based on Multiple Linear Regression and Time Series Clustering. *Mathematical Problems in Engineering*, 2020, 1–17. doi:10.1155/2020/9546792

2 Big Data-Based Time-Series Forecasting Using FbProphet for the Stock Index

Debankita Mukhopadhyay
Department of Computer Science & Technology,
Techno College of Engineering Agartala, Tripura, India

Sourabarna Roy
Department of Computer Science & Technology,
Techno College of Engineering Agartala, Tripura, India

Partha Pratim Deb
Department of Computer Science & Technology,
Techno College of Engineering Agartala, Tripura, India

*Corresponding author.

CONTENTS

2.1   INTRODUCTION

The time-series dataset has time-related information that is multipurpose for forecasting and statistical reasoning. The supermarket sales foretelling helps improve income in an enterprise surrounding. Any time-related data that is reliant upon a period-related issue can be called time-series information. In [14] Banerjee et al. noticed anomalies and irregularity on a day by day, week after week, and

DOI: 10.1201/9781003279044-2

yearly premise. If any of the above options are locked, these data types will be affected. The advancement mind plays a huge depiction in accomplishing the advantageous returns through helping store estimates. In any case, item advertisement are described by their driving, byzantine, and volatilizable nature. Henceforth, estimating lumber costs and returns is a troublesome obligation. Give or figuring of gain returns in a specific standing trade/s happens hourly. Jason Brownlee considers the significance of anticipating stem costs and their profits, scientists extrade pay as you go earth-shaking tendingto improve the fraud truth in the assertion of asset value developments and returns. In this affection, the essential assertion is that financial backers, policymakers, and business establishment's lifelessness be driving and outperforming in their immovability making in recommend sharpening the profits on their speculations. In [14] Banerjee et al. say that when gillyflower is affordable the costs fully reflect semi-public and sequestered data. Movement productivity has three structures: wan, semi-solid, and strong. The faltering state indicates that anticipated qualities can't be affected by chronicled costs. The semi-solid-state address is exposed to the transparently helpful collection. The muscular variation states that the soup cost improvement's off an end result of all sincere and floor series. Prophets Library is an open-source library for single change time-series checking. In [1] Yash Indulkar proposed that it is not difficult to utilize, intended to consequently track down the right arrangement of hyper boundaries for your model to foresee brilliantly information with standard occasional patterns and examples. On the off chance that an anticipating duplicate can cook a commendable thought of the laxation of reserve costs, then, at that point, the quality and danger involved in the resources change could be limited. It would consequently be utilitarian for financial backers and policymakers to qualify hold onto resources choices and expected measures to change the progression of interests in handle markets. Different strategies someone has been abused to guarantee they give commercial center. In [2] Ching-Hsue and Jun-He say that nevertheless, the unspecialized doubtful conditions in the have activity may replace or discontinue the accumulation industry body. Irregularity conditions could be ace by applying to expect blunder commercial center methodologies through right determining apparatuses. Cautious or presto deciding for the verification store is the fundamental provocative showcase. In [3] Guo et al. proposed that various scientists independently focused on revealing the primo determining apparatuses and strategies to get stock costs [13]. In scrutinizing periodical assessment, autoregressive unsegregated affecting standard (ARIMA) is one of the unexcelled verifiable expecting methodologies for monetary patrons to get helped and straight substance on reputation assumptions.

Time-series anticipating can be tried as various techniques and diverse hyper boundaries are accessible for every strategy. In [4] Shyi-Ming and Chao-Dian say that prophet is an additive model-based time-series data forecasting technique in which a non-linear trend is consistent with yearly, week after week, and every day occasional and occasion impacts. This turns out best for time series with solid occasional impacts and numerous seasons with recorded information. Butterflies that resist a lack of data and changing trends usually do a good job of dealing with differences. In [5] Jorge Ivan and Edission Alexander proposed that forecasting activity assumes a significant part in our routine. The reason for gauging exercises is to further develop exactness and productivity. In the field of money-related planning, it has been shown that the standard techniques for expecting the protections trade are mathematical and quantifiable models. In time-series examination, there are many time-series models like consolidated autoregressive moving ordinary (ARIMA) and summarized autoregressive unexpected heteroscedasticity (GARCH), which are used to predict stock expenses and financial market designs. In [6] SR Singh says that nonetheless, factual models for the most part gauging directly, and the factors should follow ordinary measurable dissemination for better estimating. Assuming that your overview information is introduced as phonetic scores (otherwise called "language spans" and semantic qualities for the exceptionally youthful, youthful, and old), or then again assuming your example size is tiny, then, at that point, customary determining strategies might present prescient inclination helpless outcomes. In [7] Bineet Kumar and Shilpa have proposed various predictive models based on supermarket sales to solve the problem of time series using FbProphet. The FbProphet model incorporates components of the pattern p (f), period r (f), yield y (f), and mistakes MF. In relationship studies, the ARIMA model is reasonable for transient determining, and there is still an opportunity to get better in long-haul anticipating. FbProphet couldn't accurately decide the erase point. In [8] Day and Lewis say that moreover, this concentrate simply additionally fostered the computation structure and didn't consider the effect of material components on the assumption results. In this work, the Adaptive Cubature Kalman Filtering (ACKF) estimation with ignoring factor was taken on to additionally foster FbProphet, to assist it with fitting the example and incidental plan definitively during the time-spent long stretch conjecture. In [9] the concept of Deep Learning can be complex due to its approach, which incorporates numerous calculations that can be unique about AI. It is critical to comprehend the standards as it is essential to make the right model that can address the changing qualities related to the information range. There are two motivations for learning about time series models:

- Understanding the key forces and structures that govern observable data.
- Set up your model and continue to measure, check, analyze and control.

Time -eries investigation can be isolated into two principal classifications dependent on the kinds of models that can be introduced. Two classifications are:

- Monitor model: Here the data is accumulated as yf = g (f). Assessment or discernment is seen as a component of time.
- Dynamic model: Where information is mounted as yf = g (yf-1, yf-2, yf-3) ().

In [10] there are premier longs for time-series investigation: sorting out the qualities of the peculiarity addressed through the series of perceptions, and gauging (foreseeing fate upsides of the time-series variable). FbProphet is a forecasting package deal in python that became advanced with the aid of using Facebook's information technological know-how studies team. In [11] Shyi-Ming Chen says that the bundle intends to provide enterprise customers with an effective and easy-to-use device to assist forecast enterprise effects while not having to be a professional in time collection analysis.

In [12] forecasting time collection information aims to recognize how the collection of observation will retail in the future. A time-collection record could have one or more of this kind of following component:

- Trends Component – It is the regular upward or downward motion of the statistics over the whole period.

The rest of this paper is coordinated as follows. Section 1 talks firstly about a portion of the fundamental ideas of time series determining utilizing FbProphet. Section 2 presents a SENSEX, TAIEX, and KOSDAQ anticipating technique dependent on time series. Section 3 explores the consequences of the proposed technique. The end is examined in Section 4.

2.2 METHODOLOGY

Techniques for time series evaluation can be divided into classes: repeat region strategies and time-space frameworks. The past fuse powerful appraisal and wavelet evaluation. The last choice fuses auto-relationship and cross-association evaluation. In the time-space, association and evaluation may be made in a channel-like way using scaled relationship, thusly reducing the need to perform withinside the repeat region. Also, time combination appraisal methods can be isolated into parametric and non-parametric systems. The parametric philosophies expect that the secret workspace bound stochastic procedure has a positive shape which may be portrayed by the use of a little combination of limits (for example, the use of an autoregressive or moving typical structure). In those techniques, the undertaking is to check the limits of the variation that portrays the stochastic methodology.

Then again, non-parametric approaches explicitly check the covariance or the scope of the system without tolerating that the procedure has a specific shape. Methods of time series appraisal can similarly be isolated into straight and non-direct, and univariate and multivariate.

2.3 METHODS

A stock rundown or protections trade record is a rundown that activities a monetary trade, or a piece of a monetary trade, which helps monetary patrons with working out market execution by differentiating the current level of stocks and past costs. The two rule guidelines for the document are adventure connecting with quality and straightforwardness. It advises the most effective way to set up. Monetary benefactors can place assets into a protections trade record by buying a common resource or a document store included protections trade normal resources and replicating the rundown.

When there is a disparity between the file reserve and the list store, it is known as the following blunder. A rundown of the main financial exchange files can be found in our rundown of financial exchange lists. A file is a standard method for following the presentation of a gathering of resources. Regularly, a list estimates the presentation of a container of protections attempting to reproduce a particular market region. It can be a comprehensive index covering the entire market, for example, B.Standard. Records are likewise made to gauge other monetary or financial information, for example, loan costs, expansion, or creation. Lists are regularly utilized as an action for evaluating the presence of a portfolio's return. A famous speculation procedure known as ordering is tied in with duplicating these records inactively, making an effort not to beat them. A type is a marker or proportion of something. In finance, this ordinarily insinuates the genuine assessment of changes in the insurances market. For financial business areas, stock and security market records are made from a virtual game plan of insurances that address express business areas or regions. (You can't place assets into the document).

Records are likewise made to gauge other monetary or financial information, for example, loan costs, expansion, or creation. Lists are regularly utilized as an action for evaluating the presence of a portfolio's return. A famous speculation procedure known as ordering is tied in with duplicating these records inactively, making an effort not to beat them. A type is a marker or proportion of something. This is essential for better data entry. FbProphet can be used to adjust and forecast data according to seasons, it can be used to get forecast components that can change data trends based on weekly changes, monthly changes.

$$x(f) = p(f) + y(f) + r(f) + mf \qquad (1)$$

The accompanying parts of FbProphet for fitting of dates are:

- $x(f)$ = Added Substance Regressive Model
- $p(f)$ = Pattern Factor
- $y(f)$ = Occasion Component
- $r(f)$ = Irregularity Component
- mf = Mistake Term

The market record estimates the worth of an arrangement of properties with explicit market attributes. Each list has its technique, which is determined and kept up with by the list supplier. Record strategies are generally weighted by cost or market capitalization. Different financial backers use market records to follow monetary business sectors and deal with their venture portfolios. Files are well established in the speculation of the executive's business, and the asset is utilized as a benchmark to look at execution, and administrators are utilized as the reason for making list-based common assets.

Every individual list has its technique for computing the record esteem. The numerical weighted normal is the reason for ascertaining the list since its worth is gotten from the weighted normal of the complete portfolio esteem. Hence, cost-weighted lists are more impacted by changes in the most exorbitant cost stock, while market-cap-weighted lists are generally impacted by changes in stocks, for example, the biggest, contingent upon their seriousness. Stock market indices are created by selecting specific stocks of similar companies or specific stocks that meet several predefined criteria. All of these shares are listed and traded on the stock exchange.

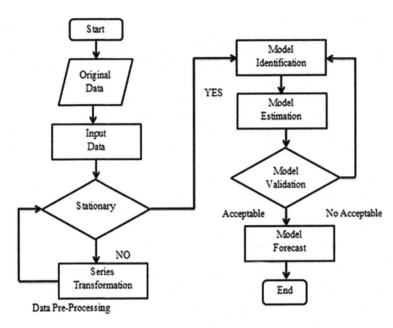

FIGURE 2.1   Flow chart of time series methodology.

Securities exchange files can be made dependent on an assortment of determination measures like industry, area, or market capitalization. Each securities exchange record estimates the value development and execution of its constituent stocks. This implies that the presentation of any securities exchange file is straightforwardly relative to the exhibition of the basic stocks that make up the file. Simply put, when the stock price of an index rises, the entire index rises. And when it fell sharply, the index fell too.

Now that we know what SENSEX is, here's how to calculate its cost: SENSEX was calculated using the free flow method. This method takes into account the percentage of easily traded stocks. The BSE then determines the free liquidity ratio after calculating the market capitalization of the 30 companies in which these shares are traded. It helps to determine the market cap in Free Explore, the ratio of the base index 100 to the stock is used to get the SENSEX value. Formula:

$$\text{SENSEX} = (\text{Complete free float market capitalization / Base market capitalization}) \times \text{Base Index} - \text{vaValue}$$

In Figure 2.1 we have shown that the insights innovation group at Facebook found that through joining mechanized determining with expert in-circle gauges for novel cases, it is feasible to cover a colossal kind of big business use-cases. We adopted Figure 2.2 from [7]. Time Series Forecasting Model for Supermarket Sales using FB-Prophet, 2021–5th International Conference on Computing Methodologies and Communication (ICCMC). FbProphet is a powerful time assortment assessment group dispatched through the way method for the Core Data Science Team at Facebook. It is

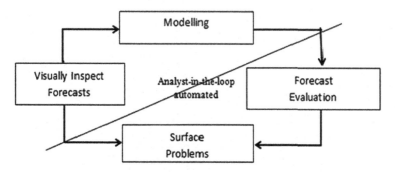

FIGURE 2.2   Forecasting process used in FbProphet.

simple and smooth to head bundle bargains for acting time assortment investigation and gauging at scale.

2.4 MATERIALS

We have taken into consideration four special types of indices for our experimental purpose. Where three records set, TAIEX (www.twse.com.tw/en/products/indices/tsec/taiex.php), KOSPI (https://finance.Yahoo.com/quote/%5EKS11/history?p=%5EKS11), and BSE-SENSEX (https://finance.yahoo.com/quote/%5EBSESN/history?p=%5EBSESN) are used for Mainfreight Ltd (MFTs). Ten months of information is used for the education section from January 1 to October 31 and two months of records from November 1 to December 31 are used for the trying out segment regardless of any strategies adopted.

2.5 EXPERIMENT RESULTS AND DISCUSSION

From the above-proposed method, we get the below graphs and the RMSE values and graphs for SENSEX, TAIEX, and KOSDAQ In this part, we notice the proposed strategy to calc the SENSEX, TAIEX, and KOSDAQ insights from 2011 to 2020. We look at the general presentation of the proposed strategy the utilization of the RMSE that is portrayed as follows:

$$\text{RMSE} = \sqrt{\frac{\sum_{i=1}^{n} (\text{forecasted value}_i - \text{actual value}_i)^2}{n}} \tag{2}$$

In which n indicates the variety of days that must be estimated. In Table 2.1, we observe the RMSE as an incentive for SENSEX, TAIEX, and KOSDAQ utilizing the worth of real anticipating and future estimating by involving the above method in condition (1).

Then in Figure 2.3, we have discussed the actual forecast for SENSEX from the year January 2011 to December 2020. In Figure 2.4 we have discussed the average

TABLE 2.1
RMSE Values and Average Values of Different Methods

Method	Year										Avg.
	2011	2012	2013	2014	2015	2016	2017	2018	2019	2020	
SENSEX	462.406	52.460	3.547	14.183	11.072	21.286	56.512	24.337	76.685	79.685	80.217
TAIEX	68.943	115.572	41.282	63.823	50.833	84.562	119.997	46.951	197.387	114.389	90.373
KOSDAQ	8.362	8.407	8.931	10.413	13.688	10.216	9.548	15.728	13.354	14.739	11.338

trend data of SENSEX for the last 10 years where the graph shows that the trend is increasing year-wise. In Figure 2.5 we have discussed the average yearly data based on actual and forecasted value, where the data is increasing and decreasing in order in every month of the year. In Figure 2.6 we have discussed the average weekly data based on actual and forecasted value, where the data is increasing and decreasing order in every day of the week.

GRAPHICAL REPRESENTATION OF SENSEX ACTUAL AND FORECASTED DATA FROM 2011 TO 2020

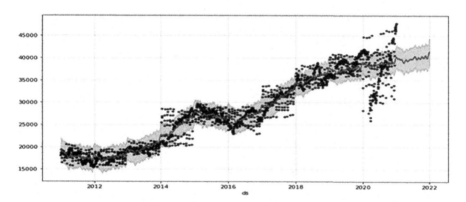

FIGURE 2.3 Actual vs forecast for SENSEX (2011–2020 & average).

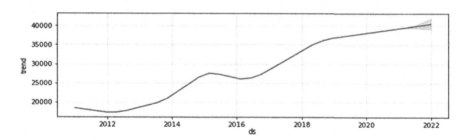

FIGURE 2.4 SENSEX actual average trend of last 10 years.

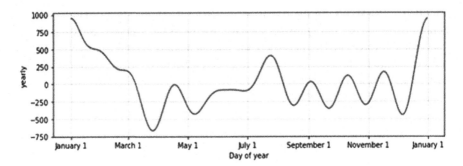

FIGURE 2.5   SENSEX actual average yearly data of last 10 years.

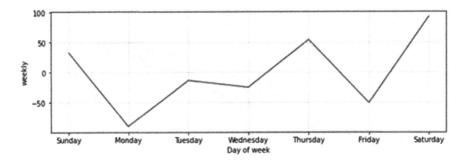

FIGURE 2.6   SENSEX average weekly data of last 10 years.

In Figure 2.7, the diagram shows the future forecast of SENSEX for the year 2020. In Figure 2.8, the diagram shows the future normal pattern esteem, where the diagram shows that the years' worth is expanding. In Figure 2.9, the diagram shows the yearly information where the chart shows that month astute information is expanding and diminishing step by step. In Figure 2.10, the graph shows the data of every day in a week where the data is increasing and decreasing form day by day.

GRAPHICAL REPRESENTATION OF ACTUAL AND FORECASTED DATA FROM 2011 TO 2020

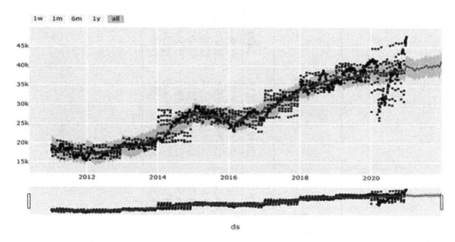

FIGURE 2.7   Future forecasting for SENSEX (2011–2020 & average).

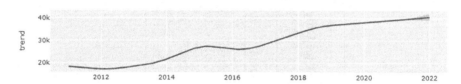

FIGURE 2.8   SENSEX future average trend data of last 10 years.

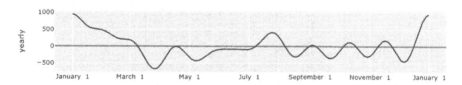

FIGURE 2.9   SENSEX future average yearly data of last 10 years.

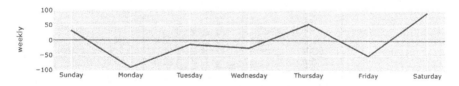

FIGURE 2.10   SENSEX future average weekly data of last 10 years.

TABLE 2.2
Comparison of the RMSE and the Average RMSE of SENSEX for Different Methods

Method	Year										Average RMSE
	2011	2012	2013	2014	2015	2016	2017	2018	2019	2020	
Huarng et al's [10]	461.625	51.465	3.402	13.567	10.674	20.316	55.437	23.543	75.473	77.598	79.310
Chen's Fuzzy Time Series [4]	463.034	47.659	1.749	12.453	10.547	20.468	54.812	21.759	76.001	74.337	78.281
Univariate Conventional Regression	459.631	50.956	2.564	12.604	10.615	18.659	54.627	24.845	75.831	77.465	78.779
Univariate Neural Network	461.026	52.021	2.785	12.764	9.965	19.456	53.706	22.744	75.102	76.497	78.606
Bivariate Conventional Regression	458.506	49.718	3.047	15.501	8.917	20.435	53.562	23.102	74.891	75.545	78.322
Bivariate Neural Network	460.321	51.094	3.491	14.022	10.901	21.015	55.072	22.781	76.849	77.018	79.256
Proposed Model	462.406	52.460	3.547	14.183	11.072	21.286	55.512	24.337	76.685	79.685	80.217

Table 2.2 indicates Comparison of RMSE values of various years along with the Average RMSE value in comparison with the earlier researches being followed with method SENSEX from the year 2011–2020.

Then in Figure 2.11, we have discussed the actual forecast for TAIEX from the year January 2011 to December 2020. In Figure 2.12, we have discussed the average trend data of TAIEX for the last 10 years where the graph shows that the trend is increasing year-wise. In Figure 2.13, we have discussed the average yearly data based on actual and forecasted value, where the data is increasing and decreasing in order in every month of the year. In Figure 2.14, we have discussed the average weekly data based on actual and forecasted value, where the data is increasing and decreasing order in every day of the week.

GRAPHICAL REPRESENTATION OF ACTUAL AND FORECASTED DATA FROM 2011 TO 2020

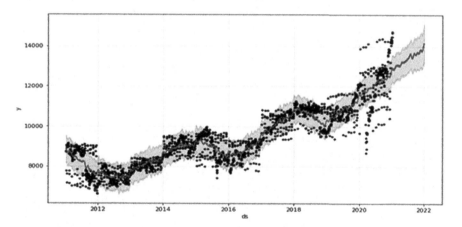

FIGURE 2.11 Actual forecast for TAIEX (2011–2020 & average).

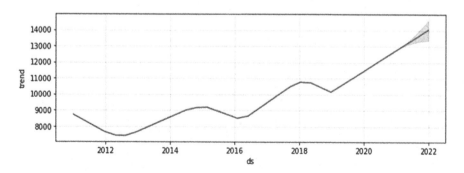

FIGURE 2.12 TAIEX actual average trend data of last 10 years.

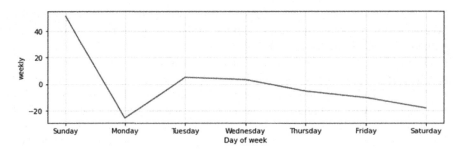

FIGURE 2.13   TAIEX actual average weekly data of last 10 years.

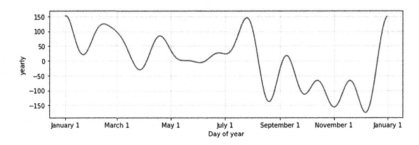

FIGURE 2.14   TAIEX actual average yearly data of last 10 years.

In Figure 2.15, the diagram shows the future estimate for TAIEX later the year 2020. In Figure 2.16, the diagram shows the future normal pattern esteem, where the diagram shows that the year's insightful worth is expanding. In Figure 2.17, the diagram shows the yearly information where the chart shows that monthly insightful information is expanding and diminishing step by step. In Figure 2.18, the graph shows the data of every day in a week where the data is increasing and decreasing form day by day.

GRAPHICAL REPRESENTATION OF ACTUAL AND FORECASTED DATA FROM 2011 TO 2020

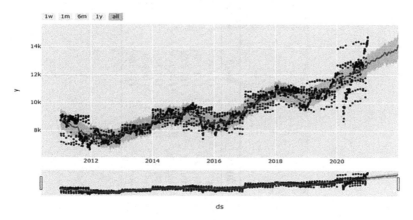

FIGURE 2.15   Future forecasting for TAIEX (2011–2020 & average).

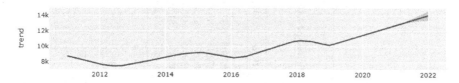

FIGURE 2.16   TAIEX future average trend data of last 10 years.

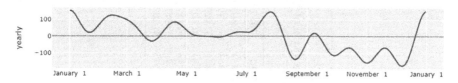

FIGURE 2.17   TAIEX future average yearly data of last 10 years.

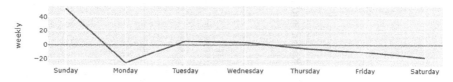

FIGURE 2.18   TAIEX future average weekly data of last 10 years.

TABLE 2.3
Comparison of the RMSE and the Average RMSE of TAIEX for Different Methods

Method	Year										Average RMSE
	2011	2012	2013	2014	2015	2016	2017	2018	2019	2020	
Huang et al's [10]	67.318	111.654	38.503	62.223	50.001	82.731	118.572	47.201	196.332	114.053	88.588
Chen's Fuzzy Time Series [4]	66.006	110.463	38.319	61.341	49.505	82.227	118.216	44.207	195.302	112.374	87.796
Univariate Conventional Regression	65.998	115.051	37.029	63.982	47.615	83.177	116.273	45.309	196.332	112.402	88.316
Univariate Neural Network	68.001	110.904	41.005	60.709	48.664	83.451	119.518	44.809	196.377	113.713	88.715
Bivariate Conventional Regression	67.871	114.762	40.611	63.006	49.641	85.011	118.541	44.772	195.031	111.709	89.095
Bivariate Neural Network	66.565	112.67	39.651	62.541	47.476	81.652	117.604	45.024	195.501	110.989	87.967
Proposed Model	68.943	115.572	41.282	63.823	50.833	84.562	119.997	46.951	197.387	114.389	90.373

Table 2.3 indicates Comparison of RMSE values of various years along with the Average RMSE value in comparison with the earlier researches being followed with method TAIEX from the year 2011–2020.

Then in Figure 2.19, we have discussed the actual forecast for KOSDAQ from the year January 2011 to December 2020. In Figure 2.20, we have discussed the average trend data of KOSDAQ of the last 10 years where the graph shows that the trend is increasing year-wise. In Figure 2.21, we have discussed the average yearly data based on actual and forecasted value, where the data is increasing and decreasing in order in every month of the year. In Figure 2.22, we have discussed the average weekly data based on actual and forecasted value, where the data is increasing and decreasing order in every day of the week.

GRAPHICAL REPRESENTATION OF ACTUAL AND FORECASTED DATA FROM 2011 TO 2020

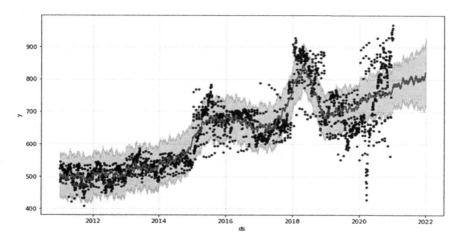

FIGURE 2.19   Actual forecast for KOSDAQ (2011–2020 & average).

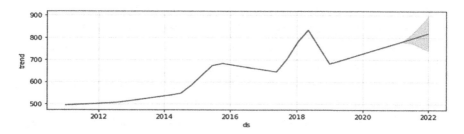

FIGURE 2.20   KOSDAQ actual average trend data of last 10 years.

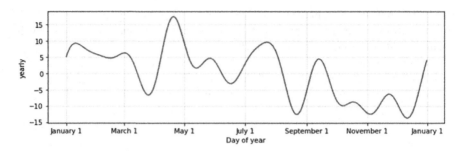

FIGURE 2.21   KOSDAQ the actual average yearly data of last 10 years.

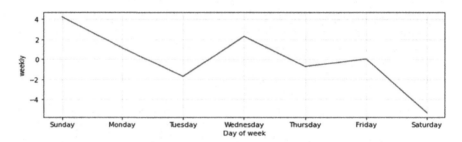

FIGURE 2.22   KOSDAQ actual average weekly data of last 10 years.

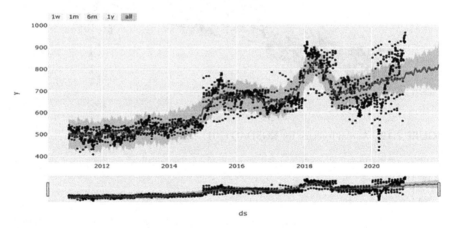

FIGURE 2.23   Future forecasting for KOSDAQ (2011–2020 & average) KOSDAQ.

In Figure 2.23, the diagram shows the future estimate for KOSDAQ later the year 2020. In Figure 2.24, the diagram shows the future normal pattern esteem, where the diagram shows that the year's insightful worth is expanding. In Figure 2.25, the diagram shows the yearly information where the chart shows that monthly insightful information is expanding and diminishing step by step. In Figure 2.26, the graph shows the data of every day in a week where the data is increasing and decreasing form day by day.

GRAPHICAL REPRESENTATION OF ACTUAL AND FORECASTED DATA FROM 2011 TO 2020

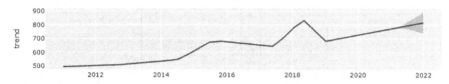

FIGURE 2.24 KOSDAQ future average trend data of last 10 years.

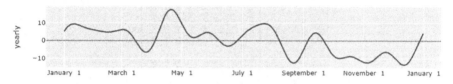

FIGURE 2.25 KOSDAQ future average yearly data of last 10 years.

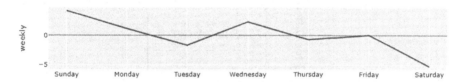

FIGURE 2.26 KOSDAQ future average weekly data of last 10 years.

Table 2.4 indicates Comparison of RMSE values of various years along with the Average RMSE value in comparison with the earlier researches being followed with method KOSDAQ from the year 2011–2020.

TABLE 2.4
Comparison of the RMSE and the Average RMSE of KOSDAQ for Different Methods

Method	Year										Average RMSE
	2011	2012	2013	2014	2015	2016	2017	2018	2019	2020	
Huarng et al's [10]	8.193	8.007	7.902	9.405	12.661	10.841	9.017	14.536	12.519	14.011	10.709
Chen's Fuzzy Time Series [4]	7.138	7.839	7.328	10.931	11.731	8.592	8.519	13.629	14.389	13.503	10.359
Univariate Conventional Regression	8.101	7.851	7.615	9.713	12.419	8.419	7.077	15.921	11.011	13.735	10.186
Univariate Neural Network	7.904	7.759	7.385	9.518	12.629	8.612	7.412	11.041	12.619	13.218	9.809
Bivariate Conventional Regression	7.924	7.703	9.002	8.735	12.629	9.003	7.343	14.339	12.593	13.551	10.282
Bivariate Neural Network	8.005	7.912	8.317	9.641	13.271	8.811	9.221	14.205	11.474	12.939	410.379
Proposed Model	8.362	8.407	8.931	10.413	13.688	10.216	9.548	15.728	13.354	14.739	11.338

2.6 CONCLUSION

This paper has proposed a big data-based time-series forecasting for the stock index, the proposed model is used to generate forecast rules. The results have shown the proposed model with better forecast performance. The prophet is easy to use, intuitive, and easy to describe the components of the model. In addition, field knowledge can be entered into the model, for example, through specific points of change or power constraints. Prediction is not bad, but in some cases, you need to tweak certain parameters from their default values, so it's easy to do. By using the above-proposed method we get the Average RMSE value for SENSEX is 80.217 TAIEX is 90.373 and KOSDAQ is 11.338.

ACKNOWLEDGMENT

We want to say thanks to Shri Partha Pratim Deb for his help during this work. We furthermore need to thank the analysts for introducing extremely helpful criticism and ideas. Their understanding and criticism achieved a higher show of the contemplations communicated on this paper.

REFERENCES

[1] Indulkar, Y. (2021). Time Series Analysis of Cryptocurrencies Using Deep Learning & Fbprophet. 2021 International Conference on Emerging Smart Computing and Informatics (ESCI). doi:10.1109/esci50559.2021.9397004

[2] Cheng, C.-H., & Yang, J.-H. (2018). Fuzzy time-series model based on rough set rule induction for forecasting stock price. *Neurocomputing*, 302, 33–45. doi:10.1016/ j.neucom.2018.04.014

[3] Guo, C., Ge, Q., Jiang, H., Yao, G., & Hua, Q. (2020). Maximum power demand prediction using FbProphet with adaptive Kalman filtering. *IEEE Access*, 1–1. doi:10.1109/ access.2020.2968101

[4] Chen, Shyi-Ming and Chen, Chao-Dian. (2010). TAIEX Forecasting Based on Fuzzy Time-Series and Fuzzy Variation Groups, *IEEE*, DOI:10.1109/TFUZZ.2010.2073712

[5] Romera-Gelvez, Jorge Ivan, Edission, Alexander Delgado-Sierra and Aurelio, Jorge. (2019). Demand Forecasting and Material Requirements Planning optimization using open-source tools, Conference: Workshops at the Second International Conference on Applied Informatics 2019, Madrid.

[6] Singh, S. R. (2007). A simple method of forecasting based on fuzzy time series, *Applied Mathematics and Computation* 186(1), 330–339, 2007.

[7] Kumar Jha, B., & Pande, S. (2021). Time Series Forecasting Model for Supermarket Sales using FB-Prophet. 2021 5th International Conference on Computing Methodologies and Communication (ICCMC). doi:10.1109/iccmc51019.2021.9418033

[8] Day, T. E., and Lewis, C. M. (1992). Stock market volatility and the information content of stock index options. *Journal of Econometrics*, 52(1-2), 267–287. doi:10.1016/ 0304-4076(92)90073-z

[9] Stock index futures Charles MS Sutcliffe Routledge, 2018.

[10] Kun-Huang Huarng; Tiffany Hui-Kuang Yu; Yu Wei Hsu (2007). *A Multivariate Heuristic Model for Fuzzy Time-Series Forecasting., 37(4),* 0–846.

[11] Chen, Shyi-Ming (1996). Forecasting enrollments based on fuzzy time series, *Fuzzy Sets Systems* 81(3), 311–319.

[12] Yu, Tiffany Hui-Kuang and Huarng, Kun-Huang (2008). A bivariate fuzzy time series model to forecast the TAIEX, *Expert Systems with Applications* 34(4), 2945–2952. doi:10.1016/j.eswa.2007.05.016.

[13] Javier, C., Rosario, E., Francisco, J. N., & Antonio, J. C. (2003). ARIMA models to predict next electricity price. *IEEE Transactions on Power Systems*, 18(3), 1014–1020.

[14] Banerjee, S., Basu, S., Nasipuri, M. (2015). Big Data Analytics and Its Prospects in Computational Proteomics. In: Mandal, J., Satapathy, S., Kumar Sanyal, M., Sarkar, P., Mukhopadhyay, A. (eds) *Information Systems Design and Intelligent Applications. Advances in Intelligent Systems and Computing*, vol 340. Springer, New Delhi.

3 The Impact of Artificial Intelligence and Big Data in the Postal Sector

Mary Olawuyi Ojo
Nasarawa State University, Nigerian Postal Service,
Abuja, Nigeria

Indranath Chatterjee
Department of Computer Engineering, Tongmyong
University, Busan, South Korea

Sheetal S. Zalte-Gaikwad
Department of Computer Science and Technology,
Shivaji University, Kolhapur, India

CONTENTS

DOI: 10.1201/9781003279044-3

3.1 BACKGROUND

This study aims to understand the impact of artificial intelligence and big data in the postal sector. In this paper, we are focusing on how artificial intelligence (AI) applications and big data will drastically impact the normal operations of the post.

The Nigerian Postal Service operating as an agency under the Federal Ministry of Communication and Digital Economy in Nigeria, has the Postmaster General as the Chief Executive Officer. The post is one of the widely distributed networks in the country, with significant offices in rural areas. It should reflect the ability to perform across various factors that enable socio-economic development.

3.2 PURPOSE AND GOALS OF THE STUDY

The impact of AI and big data cannot be over-emphasized nor exhausted. Before the fourth generation and introduction of technology, communication was standard through the post, making it a robust data home but unstructured data.

Artificial intelligence is simply the ability of a digital or computer-controlled system to perform activities commonly associated with an intelligent being. It also involves the development of digital systems and technologies that mimic human capacity, intelligence, and decision.

AI can be used in transportation, healthcare, education, customer engagement, logistics, and supply chain. Companies now use AI to enhance and automate their major processes of managing supply chains. Artificial intelligence could improve transportation efficiency and optimize performance in warehousing operations in the postal and parcel industry. By collecting and analyzing data, AI could predict inventory, materials flows, demand, supply, and other factors between business and technology (Nguyen, 2020).

"Big data" – large datasets combined with powerful analytics – is a disruptive technology trend that has the potential to change how businesses drive innovation, efficiencies, and customer satisfaction (USPF, 2014).

Big data can help unlock viable potentials, making the postal sector a major decision-making stakeholder that would drive e-government, collaborations, innovations, and e-fulfillment centers. The relevance of big data is now evidence of business value and a great asset. Technologies can now help collect, clean, extract, transform, load, and analyze unstructured postal data efficiently.

The postal sector is still touching the lives of the citizens in so many ways, especially in the rural environment despite the emerging technologies.

The rise of emerging technologies should encourage the postal sector to innovate, which contributes to the expansion and revenue generation of e-commerce services and adapt to the digital economy. We consider that if technology is indeed a factor of change, there are many different ways of adapting to it (Riot & de la Burgade, 2012). Posts need to improve their services to be more agile, responsive, relevant, and effective vehicles for socio-economic development. In particular, they need to continue expanding their reach to the most vulnerable, like the rural and underdeveloped societies. Tables 3.1 to 3.3 depict various impact of AI on the Postal Sector, in various way.

TABLE 3.1
Services Being Rendered by the Post and Yet to be Automated

S/No	Services	Functions
1.	Mail and Parcel Deliveries	The Nigerian Post (NIPOST) mandate is to collect and deliver mails six days a week and parcels five days a week at an affordable price to different post offices and destinations.
2.	Logistics	The business unit transports/conveys household, office, spare parts, agricultural produce, and manufactured goods to different locations.
3.	Financial Inclusion	**Post Transfer Domestic**: This domestic money order allows customers to send money to beneficiaries nationwide in real-time online.
		International (Remittance): The service provides means of cross-border remittance, especially within the country with MoU with the post. It also enables migrants and workers living abroad to send money back home. It is also a means of paying school fees to children or wards aboard.
		Agency Banking This allows citizens of any bank to deposit and withdraw from any bank account in Nigeria, open accounts with banks, and do other things as may be allowed by the individual banks.
		Payment Services - various utility payments such as electricity, water, subscriptions, DSTV etc - Payment of school fees.
		Mobile MONEYNIPOST serves as cash-in and cash-out outlets in all post offices. It is a payment service performed via a mobile device instead of cash, checks, or card. A customer can use the phone to pay for a wide range of services.
4.	Philately	They sell commemorative or unique postage stamps. Sales of various definitive stamps. Sales of first-day cover (FDC) Deposit account service Sales of philatelic accessories – stamp albums, magnifying lens, tweezers, etc Provide philatelic service internationally
5.	Property and Workshop	The Property and Workshop are in charge of the post's design and construction of building projects. They locate viable site/landed properties in various strategic locations nationwide for development on various Public-Private Partnership (PPP) schemes (Build, Operate and Transfer (BOT), Renovate, Operate and Transfer (ROT), etc.
6.	e-Commerce	These are responsible for online buying, selling, and shipping goods/ services to the public and government agencies.
7.	Addressing Management	The post manually verifies addresses of citizens on request for security reasons, guarantor forms, and referees. The Nigerian Post calls it Address Verification System (NIPOSTAVS).

In 2013, the International Post Corporation (IPC) and Boston Consulting Group published a report on the future perspectives of the postal sector. They indicated that building a new compelling position for incumbent postal operators requires many fundamental changes. Postal operators need to accelerate from evolutionary to revolutionary transformation to accommodate revenue decline from increased substitution and seize the window of opportunity in e-commerce (Jaag, 2015).

TABLE 3.2
Some Reviews Related to the Study

Articles	Authors	Year	Purpose
Digital Transformation of Postal Operator	Anna Otsetova	2007	To help postal sector thrive in the digital economy
The Digital Post Ecosystem Institute of Spatial Management and Socio-Economic Geography, University of Szczecin	Drab-Kurowska and Agnieszka Budziewicz-Guźlecka	2021	To address the opportunities and challenges faced by postal operators in this digital transformation age
Swiss Economics Postal Strategies in A Digital Age	Christian Jaag, Urs Trinkner and José Parra Moyano	2015	Exploring and making use of big postal data will empower postal stakeholders, enable them to take control of the future of the e-commerce and logistics sector, unveil all its untapped potentials, and reinvent postal services for the digital era.
Digital Transformation, Business Models and the Postal Industry	Prof. D. Foray, président du jury	2017	The investment in the digital platform is seen as strategic, with corresponding expectations that it will be highly profitable for the business over a longer time frame.
The Digital Economy and Digital Postal Activities	Universal Postal Union	2019	The post is a very convenient service provider for citizens because of its extensive network, including a presence in rural areas. The world is now becoming digital, businesses are becoming digital, and the effect on traditional postal services is high. Digital business means paperless trade, which in turn means that the core function of the post might change
Turning the Postal System into a Generic Digital Communication Mechanism	Randolph Y. Wang	2004	We explore the use of digital storage media transported by the postal system as a general digital communication mechanism. There are still opportunities for the postal sector in the digital age.

TABLE 3.2 (Continued)
Some Reviews Related to the Study

Articles	Authors	Year	Purpose
How Artificial Intelligence Can Affect the Postal and Parcel Industry	Nguyen, Anh	2020	AI adoption helps postal services reduce manual processing, reduce man-made errors, cut down costs. Excellent services for customers will also bring loyal relations leading to increased patronage.
Problems and Prospects of Creation of Digital Ecosystem in Postal Service O-of Uzbekistan	Z. M. Otakuziyeva Head of the Department of Technology of Postal Communication	2019	It is essential to automate the postal services with modern requirements and consumer needs, as well as for digitalization, it is necessary to improve and modernize it regularly, to introduce new technologies and to create comfortable conditions for customers.
Stampingla Poste:An Illustration of the Influence of Societal Effects on Strategic Change	Elen Riot and Emmanuel de la Burgade	2012	Strategy implementations are very necessary to move the post into a strategic place in the digital era

TABLE 3.3
The Decline in Revenue Generation

Services	Service Revenue Generation Decline (%)			
	2018	2019	2020	2021
Mails	15	20	35	25
Parcels	25	19	18	20
Logistics	15	17	10	15
Philately	5	6	4	5
Information Services	5	3	7	10
PMB	15	20	11	5
Financial Services	20	15	15	20

The above review papers consider the following research objectives:

- To know the impact of digitalization in the postal service.
- To improve the existing network and efficiency of the post.
- To ensure that the post remains relevant in the digital economy.
- To address the most critical challenges in the postal sector
  - i) letter volume decline and the,
  - ii) fast spread in e-commerce delivery volume.
- To design operational capabilities for digital services and customer interactions.

In 2020, the COVID-19 pandemic helped push traffic to the post offices, but they could not compete with other e-commerce companies due to low technology, causing a revenue drop in 2021.

E-commerce has facilitated consumers' purchases abroad and enabled enterprises to sell across borders. Overall, e-commerce has grown significantly in all EU Member States. The e-commerce markets in the EU Member States have increased their revenues from around EUR 307 billion in 2013 to EUR 602 billion in 2018. The average growth rates were 14% per year, and the experts predict that this growth will continue in the future (Otsetova, 2019).

When the systems and processes are not automated, there would be challenges delivering e-commerce services due to disjointed services, invariably making the post prone to errors and adding delays. Over the years, e-commerce trends have been entirely predictable. There was constant growth in a consumer-driven market environment. The game's rules were laid out, and the industry was playing according to those rules. And then came the pandemic, which changed everything. We have witnessed an unprecedented acceleration in e-commerce volumes due to the closure of in-store retail channels for several months and consumers' choice to order from the safety of their homes. The pandemic is still holding its grip on international commerce, and the trends that started during the pandemic will probably stay (PIP Cross-Border Report, 2021).

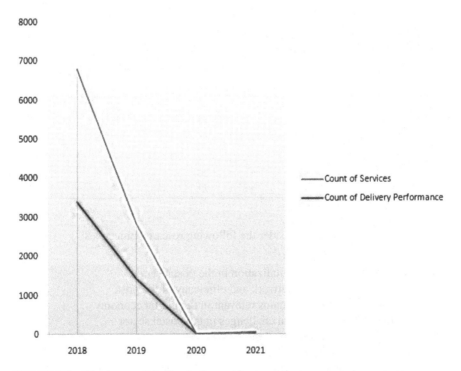

FIGURE 3.1   The degree of decline in the post in recent times.

Moreover, especially in the case of smaller items, the transfer of purchases from a physical shop to an online platform presupposes a scalable delivery network – precisely what many developing countries lack. Therefore, if e-commerce does pursue its path of growth because or despite the COVID-19 crisis, the positive externalities for the postal sector are not guaranteed, with a risk of further decline in postal relevance (UPU, 2020).

In Figure 3.1 the chart shows that the customer's performances reduced drastically, which also affected the revenue generation.

More complicated complaints, like those not getting deliveries on time, missing items, addresses not being rightly collected, can also affect the inflow of customers and traffic to the post, making it hard to maintain visibility and track the progress of the services.

3.3 DIGITALIZATION OF POSTAL SERVICES

3.3.1 The Impact of AI and Big Data in Postal Operation

With the emergence of high-performance technologies, the internet, big data, and AI have become essential for industry developments. A large amount of mail is being processed daily, and logistics processing is still in manual methods, which are time-consuming and costly to maintain. At the same time, the continuous emergence of new technologies will now have a substantial impact on every kind of profession in the world. The postal sector is more affected by digital services and e-commerce and is witnessing a drop in mail deliveries, hurting postal service revenue. AI will increase human capacities and eliminate repetitive jobs, making it easier for staff to work with more excellent added value.

The development of various digital services and e-commerce can bring many opportunities to the postal sector. Therefore, postal sectors need to digitalize and automate their system while keeping their services operational.

The postal sector faces new challenges, such as increasing competitors, new customer demands, the importance of information and communication technologies (ICT), and pressure on price rates. Postal operators have reacted positively and consider these technologies as an opportunity to modernize the postal sector (Nguyen, 2020).

3.4 THE EMERGING TECHNOLOGIES THAT ARE EXPECTED TO IMPACT THE POSTAL SERVICE THROUGH BIG DATA

3.4.1 Big Data Analytics Tools

Big data analytics tools are now making significant impacts in many industries and can also allow postal sectors to extract significant value from their existing data. Some postal operators in Europe are opening large, aggregated data sets, like national address databases or mail tracking data, to innovators or city planners so they can develop new services and applications. International postal big data analytics can provide governments with valuable real-time snapshots of Gross Domestic Product (GDP) growth, global trade patterns, and demographic movements (USPF, 2014).

The Nigerian Postal Service (NIPOST) also adopted a mini Address Verification System (AVS). The database can connect to other sources in the government to help with several seamless e-government transactions.

AVS is a GPS-based technology platform for verifying the physical address and authentication of the occupant of a claimed address using geo-tagging functionality to validate the verification exercise.

It is a post-code-compliant system to manage the processes and steps required to get addresses verified (NIPOST ABUJA, n.d.).

Postal sectors should synchronize their datasets and start collecting and analyzing data based on various institutional business objectives and by focusing on their existing data before exploring big data projects. They also need to start a transformational retreat and processes for all stakeholders and staff. Engaging senior management staff, developing proof of concept, hiring data experts, and collaborating with other agencies in the data industries to ensure success in transition.

Some necessary key agencies are the Protection and Regulatory agency, National Identity Management, National Database center, National Population Commission, all e-government departments in all agencies, and the Ministry of Communication and Digital Economy. The big data would help the National Addressing System generate many validated addresses in the system.

Over time, big data should be plugged across all the business units of the post and therefore be used as a primary decision-making tool. Various tools used are PowerBi, SQL, Tableau, SAS, MongoDB, Cloudera, Oracle, R, KNIME, RapidMiner, Hadoop etc. There are opportunities in the post with the UPU (Universal Postal Union) collaborating with postal operators when strategic steps are implemented at principal post offices.

When the Postal sector becomes a data-driven industry, it enables a convenient and cost-effective operation from collections to deliveries and successful revenue generation.

3.4.2 INTERNET OF POSTAL THINGS

The post offices scan various mails and parcels up to a hundred times daily, almost a trillion scans a year. Postal operators generate a massive amount of data from customers, staff, operations, invoices, and equipment. The physical infrastructure like mailboxes, logistics vehicles, machines, etc., has sensors connected to them to help generate data.

Posts should capitalize on the best practices of other postal sectors, private courier services, logistics companies like the DHL, FEDEX etc., to develop ideas and launch innovative big data solutions that reduce human costs, time, efforts, and generate revenue provide socio-economic growth.

For example, affixing technologically advanced sensors to parcel delivery vans/vehicles will significantly impact efficiency. They can sense the status and performance of deliveries throughout the value chain to help reduce fuel costs, fleet maintenance costs, and optimize routes.

Once the postal sector becomes an extremely data-rich industry, continuously equipping its primary network (logistics vans, mailbox, mails, and parcels, sorting stations, etc.) with sensors will expand the capability of the services.

The post has generated lots of data even before the invention of technologies, thereby giving them more information that can be used in governance and politics to make effective plans. However, some gaps should be filled to enable them to work effectively. Most postal staff are not technically inclined and have no special courses in the university related to postal technology.

It is advised that all postal operators be trained on the future job, which is the Internet of postal things. At the same time, the ICT department learns primary skills like cloud computing, robotics, programming languages to manage the environment professionally.

3.4.3 CONNECTED VEHICLES

The post can use connected vehicles and carriers with handheld devices as a platform for postal services and government agencies. This platform will support data collection, provide data collection to all government data repositories, and give the post opportunity for generating revenue to the organization and nation at large. Digital operations are essential to creating transparency around product and equipment flows and manipulating them in real-time. To enable end-to-end estimated time of arrival (ETA) transparency, for instance, incumbents need both smart assets (which track delivery vans and, increasingly, other equipment, such as roller cages) and track and trace (which tracks parcels and mail) (Briest et al. 2018).

3.5 THE EMERGING TECHNOLOGIES THAT ARE EXPECTED TO IMPACT THE POSTAL SERVICE THROUGH ARTIFICIAL INTELLIGENCE

3.5.1 LAST MILE LOGISTICS APP

The postal sector supply chain is built around these four activities: collection, transport, sorting, and delivery. These activities are traditionally labor-intensive, especially the logistics providers face many challenges and problems in overcoming the last-mile deliveries. Such challenges include strong competitive external effect on free delivery services from private companies, increased customer expectations to delivery within a short period, unscheduled delivery options, inability to track in rural areas due to a wrong network, and many unsuccessful attempts to deliver items.

Based on the research, the use of AI in last-mile coordination and delivery positively impacts last-mile delivery because AI performs great manipulation and machine-learning if properly processed well. It tracks and can analyze trends, and predicts patterns based on the findings. Delivery could be combined with an AI algorithm to determine in which order should mails or parcels be processed and delivered as fast as possible. Data from various sources can be used to optimize the route. This includes transport information, traffic patterns, GPS data, or weather information. Such data collection can significantly affect fuel, personnel, or other overhead costs associated with last-mile delivery (Jucha, 2021). Figure 3.2 depicts the supply chain of the mailing activities.

The Last Mile app solution will use AI technology to automate all registered courier routes using various parameters like the parcel delivery points, real-time

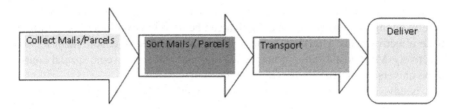

FIGURE 3.2   The supply chain of mail activity.

traffic, and weather data. The software also updates the receiver and stakeholders about the movement of all parcels.

3.5.2 AUTONOMOUS DELIVERY

These are physical and digital devices that use AI to automate the delivery task prepared by a human. Autonomous delivery does not need any human assistance during its entire process till its destination. Examples of autonomous things are flying drones, robots, self-driving cars, etc. Most drones deliver orders to challenging locations to access by dispatch riders like the mountains, hills, etc. Most customers believe it is more secure and timely.

Many delivery companies are testing and developing autonomous delivery facilities that provide logistics solutions that help improvise mobility vehicle capacity and reduce delivery time. Retailers can deliver orders to customers on the same day they receive them (Jucha, 2021).

3.5.3 OPTICAL CHARACTER RECOGNITION (OCR) MACHINES

OCR uses technology to distinguish printed or handwritten text inside a digitalized image of physical documents, e.g., scanned paper documents. These are used in mail sorting for commercial purposes. The USPS is currently using the OCR machine to sort and recognize patterns, which has aided a better mailing practice. OCR techniques have a long history in mail sorting applications. In this section, we report on first and second-generation OCR, scanning hardware, the relationship of OCR to electronic mail, performance evaluations of presently installed OCR equipment, and an on-site evaluation of an operational OCR (Hull et al., 1984). The Nigerian Post needs to adopt the OCR to generate more income. An image processing technique would help solve significant problems, such as declining postage revenue.

Mails delivered by a postal office have stamps indicating that accurate postage or actual postal permit has been processed for the items to pass through the delivery network. The newer version of OCR technologies includes some high techniques enabling address information quality. The technology can determine if an address is incomplete, thus eliminating issues with wrong delivery points, delayed delivery, and missed or lost items. Advanced OCR technologies allow for mail sorting and delivery based on country-specific address formats and languages to address the growth of a global society (Parascript, n.d.). Some uses for the OCR technology include:

1) Detecting overlapping mails to ensure that emails are not lost or sent to the wrong destination.
2) Detection of return to sender stamps, which may be detected on mail
3) To process foreign mails, differentiate between domestic and international mails.
4) To process and translate addresses written in foreign languages.

3.5.4 STAMP VERIFICATION

StampVerify is an image identification technology that detects any wrong, fake, harmful, or no postage usage on any documents. It automatically locates the stamp on all envelopes, including parcels and small packets, and verifies its identification codes on the system to ensure the validity of the information on the stamp and proper dissemination. It can generate revenue for the post by integrating the stamp verification to all tax offices and malls.

3.5.5 DOCUMENT CLASSIFICATION AUTOMATION

The technology automates and organizes documents by classifying them into various categories, either text or visual-oriented information. Text-classified documents could be incoming mails, archives, regulations, policies, bills, postcards, audit trails and data usage. The text-classified document is a language-based analysis of documents in statistics or semantic formats. The classified visual documents involve image-based documents using pattern recognition and computer vision to get insights into such documents. This technology locates and reads addresses on all parcels capturing all address information, including country, state, city, postcode, and building number. The massive amount of mail requires much more human resources to process, and the frequent international interaction adds to the burden of workers (Chung & Huang, 2022).

3.5.6 ADDRESS CHANGING AND VALIDATION PROCESS

Postal systems should create efforts to update National address databases for business use and commercial mailing orders. This can be accessible to all logistics companies, hospitals and thereby, reducing the cost of forwarding emails to wrong addresses. The post can process and initiate a change of address forms for government services to update records validate addresses in the National Address Database. Address validation technology allows businesses to update their mailing lists against their database, as this can be a significant impact on helping the postal organization generate more revenue.

3.6 CONCLUSION

AI and big data are fast becoming significant game-changers in the postal sector because all delivery and logistics companies are aware that technology is a way of making their businesses grow in this digital economy. Emerging technologies are available to turn traditional post offices into e-fulfillment service centers that fulfill citizens' expectations and introduce rural areas into the innovation process of the transaction. This is a fast and huge way to make a nation digitalized, paving more ways for a commercial environment.

The postal sector is already a huge data environment from the first century, so they do not lack data. The big postal data only needs to extract and analyze the data on their system. With the AI in postal sectors, the vehicles, post office boxes, parcels, e-fulfillment centers, etc., will help extend the opportunity of the postal industry to collect more and more valuable data. There is a severe gap in the postal with expertise as most staff were not trained to be mail carriers. The management should encourage partnerships, training, or internships with logistics companies like DHL, FEDEX, JUMIA, etc.

Postal managements need to put the necessary hands, expertise, data analysts, data scientists, and certified AI personnel to help strategize effective and efficient models to follow. This would bring together postal futurists to work together on new business solutions. They learn to manage the entire service development workflow from early incubations, analysis, proof of concepts, development, testing and the launch of the products. This workflow will bring different departments together, thereby discouraging silos solutions in the system.

The government should support the post to launch innovative solutions that reduce cost, generate national revenue, provide social and economic good to the citizens. Due to the big data, the post would also need brilliant lawyers specialized in data privacy and regulations to help postal sectors develop new policies and engagement agreements on the use of its data.

The impacts of automation of postal sectors, AI, and big data will help improve address verifications and validation, giving a company a good KPI for their customers. These improvements result in increased time-saving processes, efficiency, and reduced costs of processing government services on various platforms.

Postal services will remain a part of the critical national infrastructure. Postal services must improve and upgrade their technological infrastructure for mail processing and delivery services to maintain their mandate, ensuring that emails are delivered timely and not broken.

REFERENCES

Briest, P., Dragendorf, J., Ecker, T., Mohr, D., & Neuhaus, F. (2018). *The Endgame for Postal Networks*. McKinsey & Company.

Chung, C., & Huang, Y. (2022). The Innovation of the AI and Big Data Mail Processing System. *Journal of Research & Method in Education* 12(1), 1–8.

Hull, J. J., Krishnan, G., Palumbo, P., & Srihari, S. N. (1984). Optical Character Recognition Techniques in Mail Sorting, State University of New York at Buffalo. Department of Computer Science. Technical report 214. 1–27.

Jaag, C., Trinkner U., & Myoano, J. P. (2015). Swiss economics postal strategies in a digital age. *Swiss Economics Working Paper 0051 June 2015*, 0–14.

Jucha, P. (2021). Use of artificial intelligence in last mile delivery. *SHS Web of Conferences 92* (4011), 1–9.

Nguyen, A. (2020). How Artificial Intelligence can affect postal and parcel industry.. JAMK University of Applied Sciences, School of Technology, Finland. Bachelor's thesis,1–65.

NIPOST ABUJA. (n.d.). *AVS Marketing.*

Otsetova, A. (2019). *Postal Sector Transformation in Digital Economy. 8*(2), 110–130.

Parascript. (n.d.). *Strategic White Paper Looking Beyond the Envelope.* https://parascript.com

PIP Cross-Border Report. (2021). *PIP Cross-Border Report 2021 Report published by the Postal Innovation Platform (PIP) In February 2022.* 1–25.

Riot, E., & de la Burgade, E. (2012). Stamping La Poste: an illustration of the influence of societal effects on strategic change. *Journal of Strategy and Management, 5*(2), 175–210. https://doi.org/10.1108/17554251211222893

UPU. (2020). *Postal Development Report 2020. October.*

USPF. (2014). *International Postal Big Data: Discussion Forum Recap International Postal Big Data.*

4 Advances in Cloud Technologies and Future Trends

Indra Kumari

Division of S&T Digital Convergence, Korea Institute of
Science and Technology Information (KISTI), University of
Science and Technology (UST), Daejeon, South Korea

Jungsuk Song

Professor, Division of S&T Digital Convergence, Korea
Institute of Science and Technology Information (KISTI),
University of Science and Technology (UST), Daejeon,
South Korea

Buseung Cho

Professor, Division of S&T Digital Convergence, Korea
Institute of Science and Technology Information (KISTI),
University of Science and Technology (UST), Daejeon
South Korea

*Corresponding author.

CONTENTS

DOI: 10.1201/9781003279044-4

4.1   INTRODUCTION

"Information is cash." This expression has saturated corporate culture from Silicon Valley to Stockholm to Sydney in the last decade. Unlike stock trades, information is exchanged on and through the cloud, tying your ability to use it to your cloud abilities. Because 92 percent of organizations now use the cloud, we have compiled a list of cloud trends for 2021 and beyond to help you capitalize on this innovation. The cloud is something other than a helpful stockpiling arrangement. It is an exceptional stage for producing information and creating imaginative answers to influence that information. This extraordinary spotlight on flexibility has empowered a formerly thought-to-be-unthinkable kind of administrative reasoning. The capacity to practice has developed, permitting associations to improve their plans of action and cycles in the quest for center abilities and business objectives while staying lithe. ("Top cloud trends for 2021 and beyond | Accenture," 2021).

As we draw nearer to a cloud-based world, organizations that need to acquire an upper hand should comprehend the rhythmic movements of the cloud administration industry. We have incorporated a rundown of 25 patterns we accept that will help you contextualize your current cloud abilities and recognize regions for future development.

4.2   CLOUD COMPUTING MODELS AND SERVICES

Distributed computing advanced from the group, network, and utility processing ideas. Group and network processing utilize numerous PCs running in equal measure. Distributed computing utilizes versatile assets to address the issues of countless clients. The omnipresence of broadband and remote systems administration, falling capacity expenses, and moderate enhancements in Internet processing programming are the essential main thrusts behind distributed computing.

Cloud clients can request more assets during top jobs, decrease costs, try different things with new administrations, and eliminate overabundance limits. Multiplexing, virtualization, and dynamic asset provisioning are techniques that cloud specialist co-ops can use to build framework usage. Clients can zero in on client applications and make business sense by offloading position execution to cloud suppliers.

4.2.1 Taxonomy of the Cloud-Based on Services Provided

The headways in equipment, programming, and systems administration innovations summed up in Table 4.1 empower distributed computing. These advancements are basic to making distributed computing a reality. Most of these advances have developed in light of rising interest. In the equipment domain, quick progressions in multicenter CPUs, memory chips, and plate exhibits have empowered the development of quicker data centers with monstrous capacity limits. Fast cloud organization with HTC and catastrophe recuperation abilities is made conceivable by asset virtualization. The progression of Software-as-a-Service (SaaS), Web 2.0 guidelines, and Internet execution have all added to the ascent of cloud administrations. The present mists are designed to serve an enormous number of inhabitants while dealing with huge measures of information. The accessibility of enormous scope conveyed capacity frameworks is the bedrock of the present datacenters.

Cloud models are grouped into two sorts: administration models and sending models. Administration models are grouped depending on the sorts of cloud administrations given, though arrangement models are ordered dependent on how and by whom the cloud administrations are utilized. Cloud models are categorized into two sorts: administration models and arrangement models. Administration models are characterized as dependent on the sorts of cloud administrations given, though organization models are ordered dependent on how and by whom the cloud administrations are utilized.

4.2.2 Cloud Architecture

Distributed computing has been advanced as another processing worldview by the IT business and the scholarly community. It can give a wide scope of administrations; nonetheless, how to deal with these administrations and guarantee their quality has arisen as a basic element in the development of cloud processing. By joining the force of Service-Oriented Architecture and its idea administration the board, this paper proposes a various leveled engineering model of cloud administrations asset the executives and checking the foundation, virtualized middleware stages, and business applications as administrations.

Cloud models are ordered into two sorts: administration models and arrangement models. Administration models are ordered dependent on the sorts of cloud administrations given, though sending models are characterized as dependent on how and by whom the cloud administrations are utilized.

Be that as it may, no broadly acknowledged meaning of "cloud registration" exists. Distributed computing, which upholds the model, depends on matrix processing, virtualization, circulated and Web administrations, and service-oriented architectures (Yeo and Buyya, 2008). Cloud administrations are now utilized by organizations to work on the adaptability of their administrations and manage barges in asset interest. The cloud gives and devours virtualized actual assets. A crossover cloud is a blend of public and private mists that gives associations more adaptability. A people group or a gathering of associations deals with the local cloud.

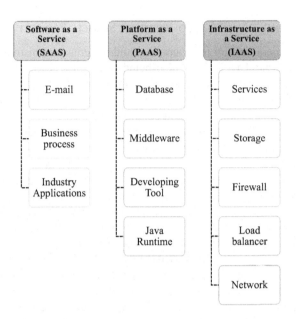

FIGURE 4.1   Service model.

Some examples of Infrastructure-as-a-service model as shown in Figure 4.1 include virtual machines and cloud storage, network components like firewalls, and configuration services. Per hour, users pay for the amount of CPU time, storage space, bandwidth, and infrastructure they use.

Applications can be deployed quickly and easily with PaaS services because there is no need to purchase or manage the underlying hardware or software or provide hosting resources.

The burden of maintaining the software is removed from customers with SaaS. The application is updated, deployed, maintained, and protected by the service provider. In the case of Gmail, for example, we are the customers. We have very limited administrative and user control, but customers can enable the priority inbox, signatures, undo send mail, and so on through the settings menu.

4.3   CREATION OF VIRTUAL MACHINES AND DOCKER CONTAINERS

4.3.1   VIRTUALIZATION

In contrast to traditional resource allocation, service workloads tend to change over time. To meet anticipated peak demand, execution environments are often provisioned with excessive resources and power, resulting in significant waste. The average utilization of resources ranges from 15% to 20% daily. The concept of virtualizing a computer's resources has been around for a long time. Virtualization improves the utilization, management, and reliability of mainframe systems.

4.3.2 Full Virtualization

It is possible to run guest operating systems in a virtualized environment. As long as the guest OS and applications are not aware of the virtualized environment, they can continue to function in the same way they would on a physical system equipped with hardware support like Intel Virtualization Technology. By using this technique, software and hardware can be completely decoupled from one another. As a result, moving applications and workloads between physical systems will be easier with full virtualization.

As a result, this technique is extremely risk-free. As part of virtualization, the VM monitor must display an image of the entire system, including the virtual BIOS and memory as well as the virtual peripheral devices. However, VM monitors, for example, are responsible for creating and maintaining a shadow memory page Table 1 for virtualized components. Every time the VMs access these data structures, they must be updated. Complete virtualization is possible with Microsoft Virtual Server and VMware ESX Server.

4.3.3 Para-Virtualization

Para-virtualization creates a hardware abstraction that is similar to but not identical to the one used by each virtual machine. Para-virtualization necessitates the modification of the guest operating systems of the virtual machines. Guest operating systems are unaware that they are running on virtual machines, which allows them to deliver near-native performance. Para-virtualization has flaws such as unauthenticated connections and data cached by the guest operating system.

Distributed computing is characterized as the development of equal processing, circulated figuring, and framework registering, just as the mix and advancement of Virtualization, Utility registering, Software-as-a-Service, Infrastructure-as-a-Service, and Platform-as-a-Service (Shahid & Sharif, 2015).

PC framework as help is given by this assistance model. This assistance is made accessible as a virtualized machine stage. Dissimilar to customary equipment machines, which require extraordinary support and have restricted adaptability, the cloud makes these machines effectively accessible basically on the web with adaptable determinations and further developed execution, streamlined to the client's necessities. Engineers can introduce and run the stages needed for programming advancement. This assistance likewise simplifies it and simple for the client to make an example for his necessary virtual machine. Most cloud administrations given by different specialist organizations take into account the arrangement of virtual machines to be finished free or for a minimal price. This virtualization include is given by the cloud as holders. For proficient virtualization, a direct virtual machine requires a hypervisor on its equipment over the bit, while containerization does not need a hypervisor, which saves processor productivity and further develops execution.

Moreover, holder size is versatile, which means it tends to be changed powerfully, which dispenses with over-provisioning. By and large, these virtual machines are introduced as plate pictures, objects, load balancers, or IP tends that can be powerfully introduced on the cloud and guarantee the security of the virtual machine by distributing a novel host address to the virtual example each time it is introduced. These virtual occasions come pre-introduced on huge pools of equipment known as server farms. These virtual machines are charged on a utility processing premise by specialist co-ops. Infrastructure-as-a-Service can give the accompanying general virtual parts:

- Computer Hardware
- Networks of computers
- Internet accessibility
- A platform virtualization environment for running virtual machines specified by the client.
- Service level agreements are number five.

4.3.4 Deployment Model

The National Institute of Standards and Technology recognizes major deployment models as Shown in Figure 4.2.

Private Cloud

This cloud sending model is overseen by a solitary association. It very well may be overseen, inside or remotely, by the association or by an outside specialist co-op. This kind of cloud is liked by organizations that have dynamic needs and require direct control over their workplace. The Amazon Private cloud by Amazon Web Services, SUSE Open Stack Private cloud, and others are instances of private mists.

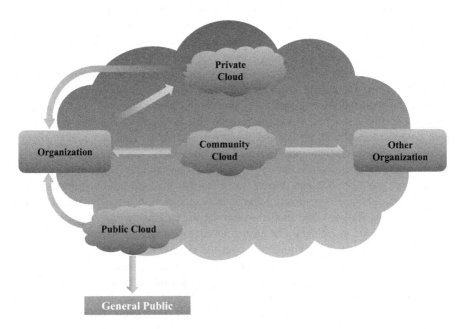

FIGURE 4.2   Deployment model.

Public Cloud

This cloud arrangement model makes its applications in general and administrations accessible and open to the public using an organization. More often than not, these administrations are free. Once in a while, the client can address his issues for practically no expense. Anybody, regardless of whether a solitary client or an association, can utilize these mists. Amazon Web Services, Google AppEngine, Windows Azure, IBM Blue cloud, and others are instances of notable public mists (Mukundha & Vidyamadhuri, 2017).

Hybrid Cloud

This sending cloud is comprised of at least two mists (public, private or local area). This model fulfills the association's security needs just like the accessibility of multi-tenure with the capacity to interface with different associations. For example, an organization might utilize a private cloud to store customer delicate information, a local area cloud to associate with different organizations, and a public cloud to interface a business insight instrument to a product application.

Community Cloud

This arrangement cloud is overseen by a local area of associations that have comparable worries, like stockpiling or security. These mists can likewise be inside and remotely facilitated and oversaw. This cloud model depends on a multi-occupant framework. Salesforce and QTS Datacenters are two local area veils of mist.

4.4   KVM AND CONTAINERS

Virtualization has arrived at an apex of significance in the field of reflection and asset the executives. Be that as it may, in a cloud climate where everything is pay-per-use, these extra layers of reflection given by virtualization include some major disadvantages as far as execution and cost.

KVM (Kernel-Based Virtual Machine) is a virtualization innovation for Linux frameworks that changes over it into a hypervisor intended for x86 processor engineering. The hypervisor depends on Linux and works on an open-source working framework. It upholds an assortment of visitor working frameworks, including Windows, BSD, and Solaris. KVM is a full virtualization procedure that permits you to give virtualized equipment like CPUs, memory, and hard circle space to the visitor working framework.

Holders offer assistance like virtual machines, yet without the overhead of running a different portion and virtualizing all equipment parts. To give reflection, it alters the host working framework. They have a holder ID and a consents framework for gatherings. Its fundamental idea depends on using piece namespaces to make detached holders. Since the interaction running inside a compartment is uninformed as far as possible, there is consistently the danger that an application will over-assign its assets (Vaughan-Nichols, 2006).

Holder security is accomplished by overseeing bunch authorizations and setting up namespace mindfulness, where clients are not allowed similar advantages inside the compartment as those external it. Docker is the most famous compartment today, conveying applications inside programming holders alongside their code, runtime, framework apparatuses, framework libraries, etc. Docker utilizes layered document framework pictures upheld by AUFS (Another UnionFS). Compartments boot up quicker than virtual machines.

The exhibition of the framework can be estimated utilizing different boundaries like throughput, inertness, transfer speed, etc. We are essentially worried about the overhead created by local, virtualized, and non-virtualized conditions. It tends to be upgraded to develop further framework execution and exhibit which one is fitting to use under favored conditions (Naveenkumar & Joshi, 2015).

4.4.1 CPU PERFORMANCE

Throughput is a boundary used to ascertain the yield of responsibility when the CPU is exposed to a pressure, High-Performance Computing (HPC) test. As can be seen, local and Docker pressure execution are equivalent, though KVM is slower.

HPC execution is equivalent on local, and Docker yet somewhat delayed on KVM because of reflection, which functions as an impediment in the present circumstance. The CPU schedulers have no impact on the processor in either the local or Docker arrangement, so there is no distinction in execution (Desai, 2016).

4.4.2 MEMORY PERFORMANCE

The boundary transfer speed is utilized to gauge the speed of memory access and activities. As indicated by different benchmarks created to pressure test memory in consecutive and arbitrary access strategies, the presentation of local, Docker, and KVM conditions is almost indistinguishable for different activities with very little deviation. The testing was done on a solitary hub with enormous datasets. Holder-based frameworks can return unused memory to the host, bringing about better memory use. Since the host and the virtual machine utilize a similar memory block, virtualization frameworks experience the ill effects of the twofold reserve.

4.4.3 NETWORK PERFORMANCE

Since the TCP/IP stack has various strategies for sending and getting information, the information move rate is estimated in two ways. The NIC is the significant part that causes a presentation bottleneck, so we measure the overhead utilizing CPU cycles. As far as execution, Docker utilizes crossing over and NAT, which stretches the course. Dockers that do not utilize NAT perform comparatively to local frameworks. KVM execution can be improved if the VM discusses straightforwardly with the host, bypassing the in the middle of layers.

4.4.4 DISK PERFORMANCE

Throughput is a measurement used to evaluate the productivity of plate activities. As recently expressed, Docker and KVM add next to no overhead when contrasted with local, yet there is a critical presentation fluctuation for KVM's situation because of a speculated bottleneck in the fiber channel. Docker has no overhead for irregular peruse and composes activities, however, KVM's exhibition endures altogether. The framework's I/O scheduler affects circle execution.

4.4.5 APPLICATION PERFORMANCE

Redis (Remote Dictionary server) is the most well-known open-source NoSQL data set utilized in compartments. It is a key-esteemed information base and an in-memory information store. Summed up customer server tasks where execution issues are just shown as organization dormancy. This is because of the customer's enormous number of simultaneous solicitations to the server. Redis is broadly utilized in distributed computing. In this test situation, the quantity of simultaneous solicitations is multiple times the number of customers. Redis is broadly utilized in the testing of systems administration and memory subsystems. Since the local climate can deal with countless customers associated with the systems administration subsystem, the number of customers associated can be handily scaled.

Since Redis is a solitary strung application, the bottleneck in execution is apparent at the CPU, which builds CPU idleness. Docker's presentation is almost indistinguishable from local, with the special case that when NAT is empowered, inactivity increments as new associations are added. KVM's presentation is lower from the outset, yet it approaches local framework execution as simultaneousness increments. Subsequently, we can reason that the Redis application should run simultaneously to be completely used.

4.4.6 APPLICATION PERFORMANCE – MYSQL

Throughput is the number of exchanges each subsequent that increments until they reach a level and become steady. Docker's exhibition is equivalent to that of local frameworks; notwithstanding, KVM has a higher overhead when contrasted with any remaining arrangements used to illustrate. The overhead presented by Docker's layered record framework is because of I/O demands being diverted through different layers. The framework's inertness increments with load, yet Docker does not show this when contrasted with different arrangements because of lower throughput at lower simultaneousness rates. Regarding CPU top use, the local framework beats Dockers, exhibiting that Dockers have a little yet huge effect (Raval et al., 2011).

When an asset is delivered, the case and all enlisted organizations are taken out from Grid ARM and GLARE. Suppose there are forthcoming solicitations for a current occurrence with the necessary organizations. In that case, the asset director can streamline provisioning by reusing a similar occasion for the following client if they share a similar cloud accreditation (or again if other trust systems permit it).

For a similar measure of CPU utilized for Docker and Docker with NAT, CPU usage estimated in throughput is insignificant; however, idleness for Docker is higher for lower simultaneousness esteems. Due to mutex conflict, MySQL can not completely use the CPU for arrangements (Dong et al. 2011).

4.5 CLOUD ARCHITECTURE AND RESOURCE MANAGEMENT

Logical registering has advanced from its beginnings with supercomputers higher than ever, for example, bunch figuring, meta-processing, and computational Grids. Today, cloud computing is arising as the worldview for the up-and-coming age of huge scope logical processing, forestalling the requirement for costly registering equipment facilitating. Researchers' Grid surroundings stay set up, and they can profit from expanding them with rented cloud assets depending on the situation. This change in outlook raises new issues that should be tended to, for example, combining this new asset class into existing conditions, asset applications, and security. The virtualization overheads for conveying and beginning a virtual machine picture are new factors when choosing booking components.

To empower the Grid climate to utilize cloud assets from different suppliers, we added three new parts to the executive's administration: the cloud the board, picture indexing, and security systems. When the superior exhibition Grid assets are drained, the scheduler can enhance them with extra ones rented from cloud suppliers to finish the work process quicker. Each cloud's accreditation properties determine the most extreme number of rented assets that can be mentioned. This limit assists with setting aside cash while remaining inside the cloud supplier's asset limits.

- Recovers a marked solicitation for the number of movement organizations needed to finish the work.
- The security part checks the solicitation's qualification and which clouds are accessible to the mentioning client.
- The picture index part recovers the predefined enrolled pictures for the available clouds.
- The pictures are verified whether they incorporate the mentioned movement sending or if they would auto be able to convey.
- In the cloud, the executive's part is utilized to begin the cases, and the picture boot process is observed until an (SSH) control associated with the new occasion is conceivable. On the off chance that the example does not have the mentioned action arrangement, a discretionary auto organization process utilizing GLARE is completed.
- In Grid ARM, another passage is made with all of the data needed by the new occurrence, like an identifier, IP address, and number of CPUs.
- The scheduler reacts from the asset chief containing the new organizations for the mentioned movement types.
- To convey a visitor working framework that shows one more conceptual and more significant level copied stage to the client, the client makes a virtual machine picture or picture. To utilize a cloud asset, the client should initially duplicate and boot a picture called a virtual machine case, curtailed as an example shown in Figure 4.3. When an example is dispatched on a cloud asset, we say that the asset has been provisioned and is prepared for use (Neto, 2011). If an asset is not generally needed, it should be decommissioned, so the client no longer pays for its utilization.

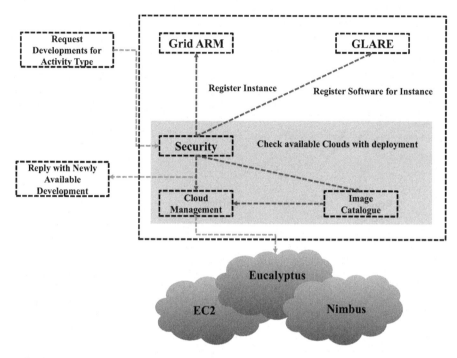

FIGURE 4.3 Cloud resource management.

To empower the climate to utilize cloud assets from numerous suppliers, we added three new parts to the asset board administration: cloud the executive's picture list, what's more, security components. At the point when the elite presentation Grid assets are drained, the scheduler can enhance them with extra ones rented from cloud suppliers to finish the work process quicker. Each cloud's accreditation properties indicate a most extreme number of rented assets that can be mentioned (Ostermann et al., 2009).

As far as usefulness, the cloud-empowered asset administrator adds two new runtime capacities to the old Grid asset supervisor: the solicitation for new arrangements for a particular movement type and the arrival of an asset after its utilization has finished. The cloud board part is responsible for provisioning, delivering, and checking an example's status.

Figure 4.4 portrays a conventional occurrence state change graph that we made after examining the occasion states in different cloud executions. At the point when a solicitation for extra assets is gotten, the cloud board part chooses the assets (occurrence types) with the best value/execution proportion and moves a picture containing the necessary action arrangements or empowered with auto-organization usefulness to them (state beginning). The picture is booted in the running state, while the case is prepared to use in the open state. The basic equipment is reconfigured during the resizing stage, for instance, by (Lenk et al., 2009) adding more centers or memory (at present just upheld by Intel).

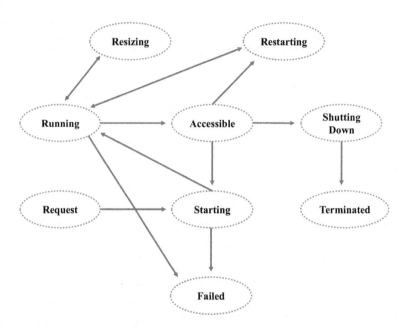

FIGURE 4.4   Cloud instance state transition.

The picture is rebooted while in the restarting stage, for instance, after a portion change. The ended state demonstrates the arrival of a picture upon closure. The bombed state indicates a mistake of any sort that makes the asset be delivered naturally. At the point when an asset is delivered, the case and all enrolled organizations are eliminated from Grid ARM and GLARE. In case there are forthcoming solicitations for a current occasion with the necessary arrangements, the asset supervisor can upgrade provisioning by reusing a similar occurrence for the following client if they share a similar cloud qualification (or then again if other trust systems permit it).

The cloud administrator additionally keeps a vault of the accessible asset classes (or occasion types) given by different cloud suppliers, including the number of centers, memory and hard circle space, I/O execution, and cost per unit of calculation. Table 4.1, for instance, contains the asset class data given by four cloud suppliers, which should be physically gone into the cloud the executive's vault by the asset supervisor director because of the absence of a comparing API.

4.6   CONCLUSION

New component technologies may lead to shifts in platform standardization, commoditization, and differentiation over the next decade. The cloud adoption curve may also be influenced by pricing and workload preferences over the decade. Security solutions will be critical in all these scenarios, as they will be critical in facilitating the widespread adoption of this fundamental paradigm shift in computing and data storage. That using autonomous computing to build smart cloud systems can improve cloud security over time is the subject of this chapter. With a variety of workloads, this

TABLE 4.1

Cloud Computing Technologies Include Hardware, Software, and Networking

Technology	Benefits and Requirements
Multi-Tenant	SaaS distributes software to a large number of users for simultaneous use and, if desired, resource sharing.
Virtual Clusters on Demand	As the workload changes, a virtualized cluster of DMs is provisioned to meet user demand.
Fast Platform Deployment	Cloud resources are deployed quickly, efficiently, and flexibly to provide users with a dynamic computing environment.
Massive Data Processing	Internet search and web services frequently necessitate massive data processing, particularly to support personalized services.
Web-Scale Communication	E-commerce, remote education, telemedicine, social networking, digital government, and digital entertainment, among other things, are all supported.
Distributed Storage	Large-scale storage of personal records and public archive data necessitates cloud-based distributed storage.
Licensing and Billing	All types of cloud services in utility computing benefit greatly from license management and billing services.

study compares Containers and Virtual Machines in terms of performance. Security solutions will be critical in all of these scenarios, as they will be critical in facilitating the widespread adoption of this fundamental paradigm shift in computing and data storage. That using autonomous computing to build smart cloud systems can improve cloud security over time is the subject of this chapter.

ACKNOWLEDGMENT

Korea Institute of Science and Technology Information (KISTI), Division of Science and Technology Digital Convergence and Intelligent Security Research Team owe thanks for the support and excellent research facilities.

CONFLICT OF INTEREST

The author declares that they have no known competing financial interests or personal relationships that could have appeared to influence the work reported in this chapter.

REFERENCES

Desai, P. R. (2016). A survey of performance comparison between virtual machines and containers. *Int. J. Comput. Sci. Eng*, 4(7), 55–59.
Dong, R., Xu, H., Gou, Y., Fu, X., & Wu, G. (2011). Analysis of land-use scenarios for urban sustainable development: a case study of Lijiang City. *International Journal of Sustainable Development & World Ecology*, 18(6), 486–491. doi: 10.1080/13504509.2011.604681.

Lenk, A., Klems, M., Nimis, J., Tai, S., & Sandholm, T. (2009, May). What's inside the Cloud? An architectural map of the Cloud landscape. In *2009 ICSE workshop on software engineering challenges of cloud computing*, 23–31. IEEE.

Mukundha, C., & Vidyamadhuri, K. (2017). Cloud computing models: a survey. *Adv. Comput. Sci. Technol.*, *10*(5), 747–761.

Naveenkumar, J., & Joshi, S. D. (2015). Evaluation of Active Storage System Realized through MobilityRPC. *International Journal of Innovative Research in Computer and Communication Engineering* 3(11), 11329–11335.

Neto, P. (2011). Demystifying cloud computing. In *Proceeding of doctoral symposium on informatics engineering 24*, 16–21.

Ostermann, S., Prodan, R., & Fahringer, T. (2009, October). Extending grids with cloud resource management for scientific computing. In *2009 10th IEEE/ACM International Conference on Grid Computing*, 42–49. IEEE.

Raval, K. S., Suryawanshi, R. S., Naveenkumar, J., & Thakore, D. M. (2011). The Anatomy of a Small-Scale Document Search Engine Tool: Incorporating a new Ranking Algorithm. *International Journal of Engineering Science and Technology*, *3*(7).

Shahid, M., & Sharif, M. (2015). Cloud Computing Security Models, Architectures, Issues and Challenges:ASurvey. *The Smart Computing Review*, 602–616. doi: 10.6029/smartcr.2015.06.010.

Top cloud trends for 2021 and beyond | Accenture. (2021). Retrieved 21 November 2021, from https://www.accenture.com/nl-en/blogs/insights/cloud-trends

Vaughan-Nichols, S. (2006). New Approach to Virtualization Is a Lightweight. *Computer*, *39*(11), 12–14. doi: 10.1109/mc.2006.393.

Yeo, C., & Buyya, R. (2008). Integrated Risk Analysis for a Commercial Computing Service in Utility Computing. *Journal Of Grid Computing*, *7*(1), 1–24. doi: 10.1007/s10723-008-9103-2.

5 Reinforcement of the Multi-Cloud Infrastructure with Edge Computing

*L. Steffina Morin
Research Scholar, Department of Computing Technologies, SRM Institute of Science and Technology, Kattankulathur, Tamil Nadu, India

P. Murali
Associate Professor, Department of Computing Technologies, SRM Institute of Science and Technology, Kattankulathur, Tamil Nadu, India

*Corresponding author.

CONTENTS

DOI: 10.1201/9781003279044-5

5.1   INTRODUCTION

During the past few years, centralized cloud computing has been regarded as a typical IT delivery platform. Despite the widespread adoption of cloud computing, new stipulations and load balancing are revealing its constraints. As a consequence of its robust data-driven paradigm, where processing and storage resources are relatively abundant and centralized, minimal or zero effort has ever been made to enhance the core hypervisor and management platform layout.

Few cloud developers have given serious consideration to the requirement of applications that require extremely high bandwidth, low latency, or a large amount of compute capacity dispersed over a number of locations, or to the requirements for supporting resource-constrained nodes that are only accessible via network connections that are unstable or have a limited bandwidth.

The development of a new infrastructure, one that is designed to directly support the global network, is required to enable new applications, services, and workloads. To meet today's requirements (retail online analytical processing, data services) and tomorrow's demands (smart cities, virtual reality/augmented reality), new scope and cloud competence requirements at remote sites are required. To keep up with changing needs, cloud maturity, resilience, flexibility, and simplicity must be extended across a growing number of sites and networks.

Despite their ease of management and flexibility, cloud computing architectures have recently been deployed to distributed infrastructures spanning multiple sites and networks. Organizations are looking for ways to incorporate cloud capabilities into WAN networks and relatively small network edge deployments. Despite the fact that distributed architectures are still in their infancy, it is becoming clear that they will benefit a wide range of new use cases and scenarios.

The purpose of this study is to look into this new requirement. Distributed cloud, fog computing, and fourth generation data centers are all terms that have been used to describe it, but for the purpose of this study, we will utilize cloud-edge computing.

5.2   CLOUD COMPUTING

Cloud computing is a large-scale, highly scalable deployment of computing and storage resources over a number of sites (regions) (Wang et al., 2020). Cloud providers also offer a number of pre-packaged services for IoT operations, making cloud deployments a popular solution. Despite the fact that cloud computing provides significantly more resources and services than traditional data centers, the nearest regional cloud facility can be hundreds of miles away from where data is collected, and connections rely on the same inconsistent internet connectivity as traditional data centers. In practice, cloud computing is utilized to replace or supplement traditional data centers in some circumstances. The cloud brings centralized processing much closer to a data source, but not at the network edge.

The three types of cloud computing services are as follows. To begin, Software as a Service (SaaS) is a web-based service that allows many users to access software as a standalone item.

Platform as a Service (PaaS) (Wang et al., 2020) is the second type of cloud computing service, which allows several operating systems to share network resources.

Infrastructure as a Service (IaaS) is the final service. It enables hardware resources to be made available as a service on demand. Elastic Compute Cloud (EC2), an Amazon service, is one of the most popular IaaS services in other nations. Amazon EC2 connects servers to provide virtualization technologies, allowing users to use the service as much as they wish.

5.3 MULTI-CLOUD COMPUTING

Multi-cloud is a strategy in which a business employs two or more cloud computing platforms to complete various tasks. Organizations who do not want to rely on a single cloud provider might pool resources from several providers to get the most out of each service. Figure 5.1 demonstrates the architecture of multi-cloud.

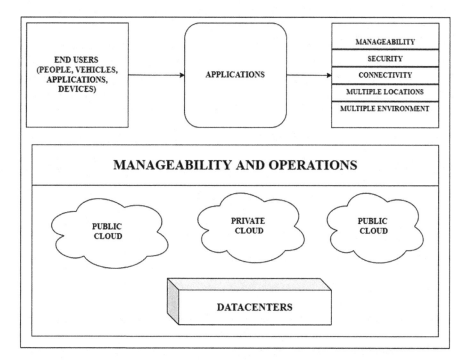

FIGURE 5.1 Multi-cloud architecture.

5.4 WHY ORGANIZATIONS CHOOSE MULTI-CLOUD

- Multi-cloud strategies are chosen by organizations for a variety of reasons.
- To reduce the risk of a localized hardware failure, organizations choose a multi-cloud strategy.
- Having to rely on a single vendor may make it difficult for a company to implement a responsive strategy.
- A failure in an on-site data center could bring the entire enterprise down.
- The use of multi-cloud reduces the likelihood of a catastrophic failure.

5.5 EDGE COMPUTING

The deployment of server and storage capabilities at the location where data are generated is referred to as edge computing. This provides data processing and storage as near as feasible to the source of data at the edge networks (Chadwick et al., 2020).

For example, a compact container with many servers and storage capacity may be positioned at a power station to receive and analyse data generated by the turbine's sensor devices.

Another example, a railway station may set aside a limited amount of compute and storage to collect and interpret data from track and train traffic sensors. Any processing results can then be sent to a separate data center for human review, indexing, and integrating with other content for more comprehensive analytics.

5.6 CLOUD EDGE COMPUTING

This is worth noting that there are numerous definitions of edge computing, some of which are contradictory. To different people, edge computing means different things. However, for our purposes, the most widely accepted definition of edge computing is that it provides application developers and service providers with cloud computing capabilities as well as an IT service environment at the network's edge. Figure 5.2 depicts the cloud-edge computing architecture.

End users and/or data inputs will be considerably more closely linked to computing, storage, and bandwidth. Edge computing environments include potentially high bandwidth across all sites, limited, unpredictable bandwidth, and unique service delivery and application functionality capabilities that a pool of centralized cloud resources in remote data centers cannot supply.

By putting part or all processing processes closer to the end user or data collection point, cloud-edge computing mitigates the disadvantages of geographically spread facilities.

In many ways, edge computing is analogous to data center computing:

- The resources for compute, storage, and networking are all provided.
- A large number of users and apps could share its resources.
- It benefits from resource pool virtualization and abstraction.
- It has the benefit of being able to run on standard hardware.
- APIs are used to facilitate interoperability.

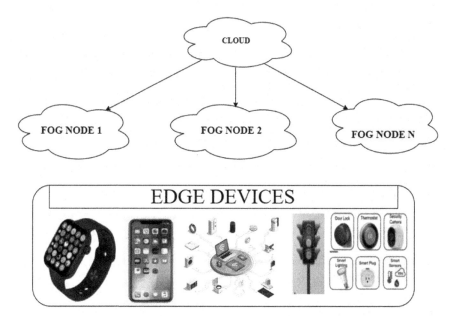

FIGURE 5.2   Cloud-edge architecture.

Computing at the edge is not as same as computing in massive data centers in the following ways:

- Edge sites are as close as feasible to end users. When dealing with high latency and unpredictable connections, they improve the user experience.
- For AR/VR capability, specific hardware, such as GPU/FPGA platforms, may be required.
- Edge can handle a huge number of sites in a variety of locales.
- The location of an edge site, as well as the identification of the access lines that terminate there, are both critical. For an application that needs to run close to its users, the right piece of the edge is essential. It is usual in edge computing for the location of the application to be important.
- The network of sites as a whole can be thought of as dynamic. WAN connections will connect the edge locations to one another and to the core in some circumstances due to their physical isolation. The infrastructure pool will experience edge locations join and exit over time.

5.7   SECURITY

New attack vectors emerge to exploit the expanding endpoint count as the number of edge devices grows—including mobile phones and IoT sensors. Edge computing provides higher-performance security applications while also increasing the number of layers that assist protect the core from breaches and danger by placing security components closer to the point of attack (Couto et al., 2018).

Protection of sensitive data in the cloud-edge environment which can be collected over the Edge devices can be done by authentication, access control, secure data storage, key provisioning, data loss prevention (DLP), and in terms of user revocation.

Security is not limited to the cloud but also with end user device security. We should be aware of the endpoint devices used by administrators to connect with the database. These edge devices should be secured, and connections from unknown or untrusted devices should be denied. Sessions should be monitored to detect suspicious activity. Some of the methods to enhance the security in the Multicloud-Edge Framework as follows.

* Distribution of Keys
* Authentication
* Access Control
* Secure Communication
* Data Loss Prevention (DLP)
* Auditing the User Accounts
* User Revocation

5.8 OPENSTACK

Openstack is a collection of open-source software projects that offer Infrastructure-as-a-Service (IaaS) through a network of interconnected services, as well as a set of software tools for building and administering public and private cloud computing systems (Thabit et al., 2021). NASA and Rackspace Hosting partnered in July 2010 to form Openstack, a cloud-software venture i.e., open-source. Openstack code was granted by NASA's Nebula framework and Rackspace's Cloud Files architecture.

Openstack is an open-source cloud operating system developed by a worldwide group of developers and cloud computing experts. For both public and private clouds, it is the most extensively used open-source cloud computing platform.

Openstack is a collection of interconnected projects that provide various components for cloud architecture and manage massive storage, compute, and networking resources across a datacenter. All of these resources are controlled using a web-based dashboard (Horizon) that allows administrators to maintain control and enabling users can provide resources.

We compared cloud computing layered models, Openstack components, and Openstack architecture for this study, as well as open-source cloud computing platforms such as Eucalyptus, Openstack, CloudStack, and OpenNebula. Finally, we will look at how to install Openstack and build virtual machines (VMs) in the Openstack cloud on Ubuntu 18.04, as well as the most current Openstack versions Victoria and Wallaby.

The Openstack project is a public and private cloud computing platform that is free and open-source. It is also Apache-licensed, which means it is free and open-source software. The cloud is mostly used to deliver remote computing services to end users, with the authenticated software being hosted as a service on stable, scalable servers instead of upon every end user's machine. Openstack enables the deployment of virtual machines (VMs) and other instances that handle various cloud management

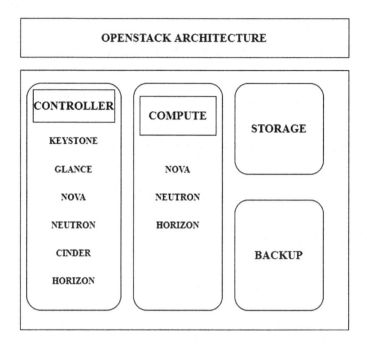

FIGURE 5.3 Openstack architecture.

functions. It enables simple horizontal scaling, allowing processes that benefit from concurrent execution to serve more or fewer customers on the fly by simply spinning up extra instances. Figure 5.3 illustrates the architecture of Openstack.

5.9 OPENSTACK COMPONENTS

5.9.1 COMPUTE (NOVA)

This is a cloud computing controller that controls and delivers vast quantities of virtual machines and other instances to handle computational tasks.

5.9.2 OBJECT STORAGE (SWIFT)

Swift is a scalable and redundant object and file storage system. Objects and files are written to several disk drives scattered among data center servers, with Openstack software managing data redundancy and reliability.

5.9.3 BLOCK STORAGE (CINDER)

Cinder is the compute instance block storage component that provides permanent block-level storage devices. It is similar to the idea of a computer being able to access specific parts of a disk drive in the past. Adding, removing, and connecting block devices to servers is handled by the Openstack block storage system.

5.9.4 Networking (Neutron)

Neutron is a framework for easily, swiftly, and efficiently managing networks and IP addresses. Figure 5.4 shows how the neutron works.

5.9.5 Dashboard (Horizon)

Horizon is a graphical user interface for accessing, provisioning and automating cloud-based resources for administrators and users.

5.9.6 Identity (Keystone)

Keystone is a pivotal list of users that is linked to the Openstack services they can utilize. It allows many access points, acts as the cloud operating system's standard authentication mechanism, and connects with extant backend directory services like LDAP.

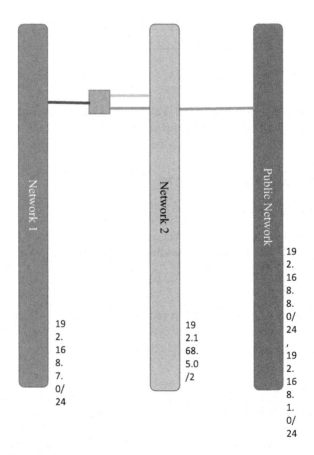

FIGURE 5.4   Openstack network.

5.9.7   Image Service (Glance)

Glance offers image services such as disk and server image discovery, registration, and delivery, as well as the ability to utilize these images as templates for creating new virtual machine instances.

5.10   REINFORCEMENT OF CLOUD-EDGE INFRASTRUCTURE WITH OPENSTACK

5.10.1   Creation of Personal Private Multi-Cloud-Edge Infrastructure using Openstack

Thus, the creation of own private multi-cloud-edge infrastructure using Openstack with the features as follows.

5.10.2   User Authentication

Authentication (Namasudra, 2019) is the systematic verification of user's identification—that they are who they say they are. Entering a username and password while logging onto a system is a well-known example.

Authentication methods supported by Openstack Keystone include user name and password, LDAP, and external authentication (Mondal and Goswami, 2021). The Identity service sends the user an authorization token after successful authentication, which can be used for subsequent service requests.

5.10.3   Access Control

Most Openstack services use Role-Based Access Control (RBAC) to control user access to resources (Esposito et al., 2017). If a user has the necessary role to perform an action, authorization is granted. Figure 5.5 shows the entity and their roles depicted.

5.10.4   Data Loss Prevention (DLP)

Cloud DLP gives you access to a powerful platform for sensitive data inspection, classification, and de-identification.

FIGURE 5.5   Openstack access control.

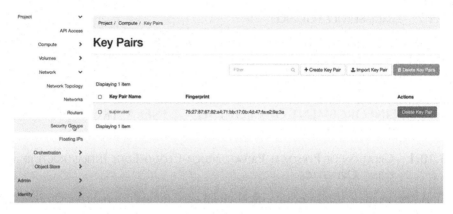

FIGURE 5.6   Openstack security with key pairs.

5.10.5   MONITORING THE USER ACCOUNTS

Using Applications Manager to monitor Openstack ensures that you always have access to all of the necessary information about your Openstack deployment, including performance, health, and availability statistics (Yu et al., 2019).

Receiving instant alerts when your Openstack server's performance is degraded and take corrective action as soon as possible. Create dynamic baselines to detect gradual performance degradation and make informed resource allocation and capacity planning decisions.

5.10.6   SECURE THE DATA IN STORAGE AS WELL AS DATA-IN-TRANSIT

We have designed the robust network security authentication methods to secure the data in storage as well as data-in-transit in Openstack. Figure 5.6 represents the key pairs for the Openstack security.

5.10.7   USER REVOCATION

In Openstack, token revocation is used to prevent a token from being used again. In certain circumstances, such as when a user logs out of the Dashboard (Anilkumar and Subramanian, 2021), the token is immediately revoked. If another party obtains a revoked token, it should no longer be allowed to use it to execute any cloud actions.

5.11   RESULTS AND DISCUSSION

Security is not limited to the cloud but also with end user device security. We should be aware of the endpoint devices used by administrators to connect with the database. These edge devices should be secured, and connections from unknown or untrusted devices should be denied (Esposito et al., 2017). Sessions should be monitored to detect suspicious activity.

5.12   CONCLUSION AND FUTURE WORK

Hence, the process of limiting potential weaknesses makes the multi-cloud-edge infrastructure with Openstack vulnerable to cyber attacks. Complex security needs at the cloud's edge entail the provision of a large number of basic services.

Future work in this area can be done by introducing new light weight cryptographic schemes with the infrastructure to reduce the computational complexity of cloud and edge-based applications.

REFERENCES

Anilkumar, C., & Subramanian, S. (2021). A novel predicate based access control scheme for cloud environment using open stack swift storage. *Peer-to-Peer Networking and Applications*, *14*(4), 2372–2384.

Chadwick, D. W., Fan, W., Costantino, G., De Lemos, R., Di Cerbo, F., Herwono, I., Manea, M., Mori, P., Sajjad, A., & Wang, X. S. (2020). A cloud-edge based data security architecture for sharing and analysing cyber threat information. *Future Generation Computer Systems*, *102*, 710–722.

Couto, R. S., Sadok, H., Cruz, P., da Silva, F. F., Sciammarella, T., Campista, M. E. M., Costa, L. H., Velloso, P. B., & Rubinstein, M. G. (2018). Building an IaaS cloud with droplets: a collaborative experience with OpenStack. *Journal of Network and Computer Applications*, *117*, 59–71.

Esposito, C., Castiglione, A., Pop, F., & Choo, K. K. R. (2017). Challenges of connecting edge and cloud computing: A security and forensic perspective. *IEEE Cloud Computing*, *4*(2), 13–17.

Mondal, A., & Goswami, R. T. (2021). Enhanced Honeypot cryptographic scheme and privacy preservation for an effective prediction in cloud security. *Microprocessors and Microsystems*, *81*, 103719.

Namasudra, S. (2019). An improved attribute-based encryption technique towards the data security in cloud computing. *Concurrency and Computation: Practice and Experience*, *31*(3), e4364.

Thabit, F., Alhomdy, S., Al-Ahdal, A. H., & Jagtap, S. (2021). A new lightweight cryptographic algorithm for enhancing data security in cloud computing. *Global Transitions Proceedings*, *2*(1), 91–99.

Wang, L., Yang, Z., & Song, X. (2020). SHAMC: A Secure and highly available database system in multi-cloud environment. *Future Generation Computer Systems*, *105*, 873–883.

Yu, Y., Chen, R., Li, H., Li, Y., & Tian, A. (2019). Toward data security in edge intelligent IIoT. *IEEE Network*, *33*(5), 20–26.

6 Study and Investigation of PKI-Based Blockchain Infrastructure

Rashmi Deshmukh
Assistant Professor, Department of Technology,
Shivaji University, Maharashtra, India

Sheetal S. Zalte-Gaikwad
Assistant Professor, Department of Computer Science and
Technology, Shivaji University, Maharashtra, India

*Corresponding author.

CONTENTS

6.1   BACKGROUND

In the present, mostly connected environment, most of the web-borne dangers are from attackers on the frameworks and information. Hence, public key infrastructure of the system should hold the keys and must be secure enough.

Public Key Infrastructure (PKI) provides secured communication between client and server. The digital certificate gives protection to the user data and communicates securely with a server, but the drawback of PKI is that this certificate issuing authority is given to another website, called Certificate Authorities (CAs). If the CA systems are attacked by an unsecured system, then this problem cannot be solved easily. These drawbacks of centralized handling of CAs by PKI can be handled by blockchain.

DOI: 10.1201/9781003279044-6

6.1.1   PKI

A PKI is a framework for making and distributing digital certificates to confirm whether a particular public key has a place with a specific organization. The PKI innovation depends on a mix of private and public keys so somebody who has the private key can decode encoded messages with the public key. A PKI framework is asymmetric as the client should get to both the public key and the beneficiary's private key to decrypt data (Remme, 2019a, b).

Public key infrastructure includes digital certificates and cryptographic keys which give secure connections for user and machine identities. It offers the roles and policies for creating and managing the revoked digital certificates, and also encrypts public keys. A survey by Pulse Research and Key Factor finds priorities and challenges of enterprise security for zero-trust strategies and the implementation and use of PKI and digital certificates in zero-trust architectures. (Wikipedia, n.d.; Pulse Research and Keyfactor, 2021).

A PKI is secured, as it is used to verify users and servers over a large network. It is also used to establish a secure digital world where a digital certificate acts as authenticated digital information. The certificate is used to confirm possession of the public key by restoring it along with information about the owner and some organizational data. Public key infrastructure technology uses private key for decryption, and encryption is done with the public key. Secured lock signatures are used while accessing any information from authenticated sites. Safety is checked and maintained with the digital certificate of a page.

Figure 6.1 shows the architecture of the traditional PKI system. Digital certificates create a key pair for binding a specific user to a certificate. After this the system verifies the user's identity and secure connections are made with any website or browser (Remme, 2019a, b).

6.1.2   Problems with PKI

The public key infrastructure is centralized, as only certificate authorities can issue universally valid certificates. Public key infrastructure works on digital certificates, so security problems are related to trusted third parties called CAs. Problems with CAs include giving certificates to unauthenticated users, no immediate revocation of the certificate, wrong use of the public and private keys, etc. These are due to bad certificate management practices.

6.1.3   What is Blockchain?

The idea of the blockchain was first presented in the Bitcoin framework. Blockchain mainly contains the distribution of data and updating of data done by each participating node, called a block, in a network. The data is stored in the form of replication and further shared across these nodes.

The difference between PKI and blockchain is informing the true identity of participants. Whilst PKI finds a specific user who is authenticated or registered, the public key of a blockchain has not been identified by any authorized system.

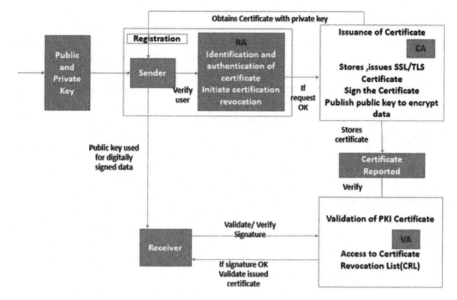

FIGURE 6.1   PKI architecture.

Blockchain is a novel method that shares information without the involvement of third parties and additional transaction costs. The problems of traditional PKIs are overcome by blockchain technology with provides a fast response to CAs deficiencies (Gisolfi, 2018).

The blockchain contains thousands of computers designed to eliminate the risks found in traditional PKI systems. Key Factors Crypto Agility Platform enables security teams to orchestrate keys and certificates for their entire organization. Key Factor is a leader in cloud-first PKI services and crypto agility solutions (Hackernoon, 2019).

Blockchain focuses on regulated environments such as securities trading, and intelligent contracts are processed by a network of computers used in blockchain. Thus, blockchain provides trust through smart contracts, and transaction processing capacity is used as a skeleton (Remme, 2019a, b).

Blockchains are able to back up the secured data and lower the involvement of third parties or any unauthenticated entity in the system. This solution protects information in a secure and distributed manner (Momot, 2019).

Blockchain is a ledger of distributed databases of transactions containing a list of records in the form of blocks. All entries are linked to each other. Blockchain is decentralized, hence it is operated by all members of the system. Blockchain maintains robust transaction records so events recorded in the past cannot be modified without permission of the majority of blocks connected in the network. This is done through smart contracts, reputation systems, and IoT (Internet of Things) device interactions. Hence blockchain is more secure compared to traditional PKI.

6.1.4 PKI with Blockchain

Decentralized PKIs with blockchain give certificate transparency by public logging and monitoring of certificates. Blockchain also eliminates points-of-failure occurred by certification authorities with the use of chains of certificates.

Public key infrastructure provides secured communication between client and server. Two keys (public and private) protect the browsers' or clients' information from unsecured websites. The digital certificate gives protection to the user data and communicates securely with a server. However, the drawback of PKI is that this certificate issuing authority is given to another website called a certificate authority (CA). If the CA systems are attacked by an unsecured system, then this problem cannot be solved easily. Bad certificate management practices also cause a problem in the secured system. These drawbacks of centralized handling of certificate authorities by PKI can be handled by blockchain. Blockchain architecture uses distributed databases by maintaining data in the form of blocks. This architecture is secured as a large set of computers working in parallel. The problem of relying on CA is eliminated with this. In the blockchain, any peer node can read the context of the block. During the issue of the key, if any problem occurs then this information is known to all peer nodes in the chain. If anyone changes any item from a block, then all items need to be changed as all blocks are connected. Figure 6.2 shows an architecture of the PKI system with blockchain.

A distributed database is maintained by blockchain for storing Personal Identifiable Information. These are stored with a distributed hash table. With blockchain, the client and server maintain their public key and decentralized identifier on a distributed public ledger. With a public key, any peer node can access the information. Blockchain records digital information and distributes it but does not allow easy updating.

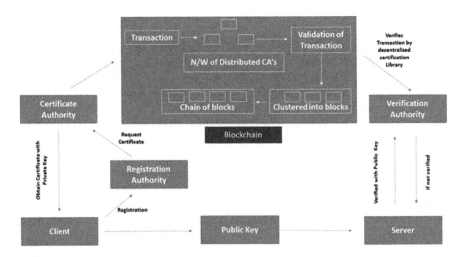

FIGURE 6.2   PKI with blockchain architecture.

6.2   NECESSITY OF PKI WITH BLOCKCHAIN

- Traditional PKI is having problems related to issuing Certificate Revocation Lists (CRLs). A long-standing problem is solved by blockchain with revocation checking on the local copy of the user or verifier. It also automatically responds to the misbehavior of certificate authorities. Also, incentives are given to these nodes who can detect these misbehaviors.
- Blockchain uses decentralized public key infrastructure where every identity is by its principal owner. Secured connections are formed by providing SSL certificates.
- Blockchain maintains smart contracts for trusted transaction processing. As per the requirement, transactions are processed by smart certificates and all nodes in the network actively participate in the processing of smart contracts. This process does not contain any manual work, and hence it is more reliable.
- Blockchain provides transparency, as all nodes have access to smart contracts; and integrity, as no one can do the changes to the system, which ensures interoperability with reduced resources; and more availability, with a distributed network of nodes.
- Blockchain is more robust and durable. With all nodes on the network processing transactions, updating of the records and errors in calculations are minimized (Remme, 2019a, b).

6.3   LITERATURE SURVEY

The following section presents a literature survey of the research papers showing the advantages of PKI with blockchain infrastructure.

The traditional PKI model does not contain transparency hence it is hard to detect targeted implementation threats. Boyen (2021) explains that Decentralized PKI Transparency (DPKIT) maintains transparency during the issue of a certificate and removes a single part of failure. With this, any browser can check the history of public certificates. This certificate provides transparency with a security model for append-only associative ledgers.

The drawback of Google's certificate transparency is shown in the paper as a certificate which lacks consistency for domains having multiple certificates. This mechanism for rewriting certificates can lead to a single-point failure. The paper focuses on an algorithm for implementation among peer nodes and the DPKIT structure containing certificate authorities, clients, servers, and domain owners.

Talamo (2020) has proposed a solution for weaknesses with PKI in a non-hierarchical context for the consensus algorithm developed for blockchain technology. The author has used a distributed approach with trust for the certificate expressed by a set of distributed users.

The X.509 certificate is trust managed with the use of distributed ledger technology. This approach uses checking of the certificates by all peer nodes and approval for the same is taken from all peer nodes and maintained.

The paper explains the consensus algorithm with smart contracts activated by blockchain peers. It validates the suite and its certificate. Further blockchain management executes a smart contract with this consensus algorithm and randomly selects a peer for the validation. If the algorithm gets the same results, an algorithm updates the respective blockchain and sends updates to all other peer nodes. And if results are not identical from the majority of peer nodes it is considered as a fake and a new peer is selected for validation of the smart contract.

The paper highlights the major advantages of the consensus algorithm. An algorithm is more adaptive and also has less response time for consensus. This algorithm detects MIM (Man in the Middle) attack and gives more protection to the client having less and not proper information.

The paper presents an extension field for X.509 certificate and this gives the details of peer node with revocation of the certificate. The paper presents this mechanism of public blockchain with Python and the Namecoin blockchain.

The paper highlights on revocation status information structure with new hashed revoked certificate called bloom filter and re-evaluated with adding a newly revoked certificate. Light Revocation Structure Information (LRSI) uses the detection of false-positive information. Also, LRSI saves bandwidth and transaction time.

Experimental work represented by Elloh Yves Christian et al. (2021) shows a stable time required with a Bloom filter for an approach. The method outperforms earlier methods such as Online Certificate Status Protocol (OCSP) and certificate revocation tree. Also, this method performs well compared to CRL as less data is needed for verification of revocation status. With a public blockchain, the method becomes more scalable. With the use of LRSI, time for communication with certificate authorization is decreased. Unlike the OCSP method, this approach can access data without the Intent method with LRSI having no exposure of information. Hence, the approach is more secure and fully compatible with web standards of the current situation. It uses a more robust decentralized blockchain architecture and is protected with blockchain design constraints with signed LRSI (Lightweight revocation status information) at the client end.

Khieu (2020) proposes cloud-based public key infrastructure (PKI) with blockchain technology. Traditional PKI is not suitable for big data ecosystems, and hence with the use of blockchains it is possible to access certificate data and CRLs, hosting it to cloud providers. Certificate data and status are combined into smart contracts. Thus cloud-based PKI (CBPKI) enhances scalability and forms a persistent identity management system. The method assumes that hash functions, digital signatures are secure, and deployed dynamic accumulators are correct.

For any malicious activities, any node of the system can audit and start revocation. Revocation lists are also not needed as any node is able to verify the public keys through witnesses of blockchain. Effective verification of the public keys done with dynamic Merkle tree-based accumulator. Trust is distributed through Web of Trust and enrolment of new keys is done with a consensus mechanism between trusted nodes. Enrolment, revocation, and update procedures are more secure as it uses the Practical Byzantine Fault Tolerance consensus mechanism and the deployed accumulator Toorani (2021).

Blockchain overcomes problems associated with PKI with the use of public ledgers. Axon (2017) presents a Privacy-aware Blockchain-based PKI (PB-PKI). As blockchain is controlled in a decentralized manner hence can provide a more secure system. The key update process in PB-PKI is different where a public link is not given. The key is formed instantly by a combination of initial identification and an updated key. This forms a hidden link, and this allows the user to update his public key anonymously.

This paper presents the Smart Contract-based PKI and Identity system (SCPKI) as a system hosted on the Ethereum blockchain and controlled by a smart contract. This store retrieves and verifies identities of itself and another node within a network. The SCPKI maintains a smart contract for setting the protocols. Al-Bassam (2017) shows a decentralized public key infrastructure system that utilizes the transparency of the blockchain.

6.4 OBJECTIVE

- To study different PKI-based blockchain architectures.
- To analyze and investigate more robust and faster architecture for secured digital information.

6.5 METHODOLOGY

We have studied blockchain architecture in detail. Blockchain architecture uses distributed databases by maintaining data in the form of blocks. This architecture is secured, as a large set of computers are working in parallel. A distributed database is maintained by blockchain for storing Personal Identifiable Information. These are stored with a distributed hash table. With blockchain, the client and server maintain their public key and decentralized identifier on a distributed public ledger. With a public key, any peer node can access the information. Blockchain records digital information and distributes it and is not allowed easily to update.

This paper gives robust PKI-based on blockchain framework to solve the problem of a single point of failure problem with traditional PKI. This solution is given with two methods as log-based PKI schema and web of trust. With log-based PKI schema, only publicly logged certificates are validated, and a single point of failure drawback is overcome with the distributed approach of web of trust where the certificate is trusted by the third party. The advantages of these two approaches are used by blockchain-based PKI which gives a more secure PKI system.

Yakubov (2018) mostly focuses on designing hybrid x.509 certificates. The hybrid certificate contains extensions fields like subject by identifier, blockchain name, CA key identifier, issuer CA identifier, hashing algorithm, etc. This framework contains a hierarchy of certificates that are linked to each other with CA contract ID to issue CA ID. This hierarchical framework of blockchain-based PKI contains subparts as a restful source, certificate validation, and web user interface for testing. Experimentation in the paper shows that REST source performs faster than traditional PKI's smart contract as in blockchain framework any updating is immediately get shared with all nodes. Also, it gives more protection to MITM attacks as blockchain works decentralized and revoking of certificate information is sent to all peer nodes.

6.6 RESULTS AND DISCUSSION

The proposed study theoretically analyses robust PKI based on blockchain framework to solve the problem of a single point of failure problem with traditional PKI. The solution is given with two methods as log-based PKI schema and web of trust. Certificate revocation with blockchain presents an extension field for the X.509 certificate and gives the details of peer node with revocation of the certificate. The DPKIT system gives more transparency and distributed network for the PKI system. Testing of issued certificates and revocation of the same is done effectively with DPKIT. The DPKIT presents a secured model for maintaining certificate transparency with associative ledgers with the read-only facility (Boyen, 2021).

6.7 CONCLUSION AND FUTURE WORK

Blockchain is an inventive technology that enables information to be shared without third parties and without transaction costs. Blockchains are able to back up the secured data and also lower the involvement of any unauthenticated members in the system. This protects information in a secure and distributed manner. Blockchain provides more transparency through smart contracts, maintains integrity, and gives a more robust and durable system. Smart contract-based PKI and Identity system and Privacy-aware Blockchain-based PKI (PB-PKI) are exclusive blockchain-based PKI systems. Certificate revocation system with public blockchain system uses a data structure bloom filter to find the unknown node from the network which offers distributed revocation and verification of the certificates. Studies show that PKI with blockchain performs better than previous approaches like CRL and OCSP. Hence PKI with the use of blockchain technology offers a more secure system in modern cryptocurrencies.

REFERENCES

Al-Bassam, Mustafa (2017). SCPKI: A Smart Contract-based PKI and Identity System. ACM ISBN 978-1-4503-4974-1/17/04, DOI: http://dx.doi.org/10.1145/3055518.3055530

Axon, Louise and Michael Goldsmith (2017). PB-PKI: a Privacy-Aware Blockchain-Based PK. https://www.researchgate.net/publication/318870515. DOI:10.5220/0006419203110318.

Boyen, Xavier et al. (2021). Associative Blockchain for Decentralized PKI Transparency. Cryptography EISSN 2410-387X. Published by MDPI. https://doi.org/10.3390/cryptography5020014.

Elloh Yves Christian, A. et al. (2021). A blockchain-based certificate revocation management and status verification system. *Computers and Security* 104. DOI: 10.1016/j.cose.2021.102209.

Gisolfi, Dan (2018). Self-sovereign identity. Why blockchain? *IBM blog* [online] Available https://www.ibm.com/blogs/blockchain/2018/06/self-sovereign-identity-why-blockchain/

Hackernoon. (2019). What is it and why does it matter? *Hackernoon* [online]. Available at https://hackernoon.com/decentralized-public-key-infrastructure-dpki-what-is-it-and-why-does-it-matter-babee9d88579

Khieu, Brian Tuan and Melody Moh (2020). Cloud-Centric Blockchain Public Key Infrastructure for Big Data Applications. DOI: 10.4018/978-1-5225-9742-1.ch005.

Momot, Alex. (2019). How Blockchain Addresses Public Key Infrastructure Shortcomings. *Infosecurity Magazine* [online]. Available at https://www.infosecurity-magazine.com/opinions/blockchain-pki-shortcomings-1-1/

Pulse Research and Keyfactor (2021). [Press Release] Survey Findings from Pulse Research and Keyfactor Show Gap Regarding PKI's Role in a Zero Trust Security Strategy. Available at https://www.pr.com/press-release/832829

Remme. (2019a). How blockchain addresses public key infrastructure shortcomings. *Remme's blog* [online]. Available at https://remme.io/blog/how-blockchain-addresses-public-key-infrastructure-shortcomings

Remme. (2019b). Why next generation PKI will reside on the blockchain. *Remme's blog* [online]. Available at https://medium.com/remme/why-next-generation-pki-will-reside-on-the-blockchain-44befcd2af3a

Talamo, Maurizio (2020). A Blockchain based PKI Validation System based on Rare Events Management. *Future Internet* 12(2), 40. EISSN 1999-5903. Published by MDPI. https://doi.org/10.3390/fi12020040.

Toorani, Mohsen and Christian Gehrmann (2021). A Decentralized Dynamic PKI based on Blockchain. https://doi.org/10.1145/3412841.3442038.

Wikipedia. (n.d.) Public key infrastructure. *Wikipedia* [online]. Available at https://en.wikipedia.org/wiki/Public_key_infrastructure

Yakubov, Alexander et al. (2018). A Blockchain-Based PKI Management Framework. IEEE/IFIP man2block 2018 At: Taipei, Taiwan: 10.1109/NOMS.2018.8406325.

7 Stock Index Forecasting Using Stacked Long Short-Term Memory (LSTM)
Deep Learning and Big Data

*Debasmita Pal
Department of Computer Science & Engineering,
Techno College of Engineering Agartala

Partha Pratim Deb
Department of Computer Science & Engineering,
Techno College of Engineering Agartala

*Corresponding author.

CONTENTS

7.1   INTRODUCTION

The stock market prediction research has long been a major attraction for investors and financial institutions. The successful stock market forecast is constantly supporting people in making an early decision to buy or sell stock market shares. Such early decision allows them to get the most out of their investment. These financial time-series data are more complex than other time-series data because they are significantly more dynamic and have a long-term dependency between them. Predicting these highly fluctuating and irregular datasets is a difficult task. There are several complex financial indicators, and stock market fluctuations are quite volatile. Meanwhile, as technology advances, the chance to generate a consistent profit from the stock market grows, and it also helps specialists in identifying the most useful indicators for making better predictions. The ability to estimate market value is important for maximizing profit from stock option purchases while minimizing risk. These are challenging tasks in

DOI: 10.1201/9781003279044-7

research (Barak and Modarres, 2015). To construct trading strategies based on financial modelling, it is very important for the model to be able to learn the pattern in the time-series data and provide accurate predictions (Kim and Han, 2000). Over the years, many different research studies applied different techniques on the analysis of stock predictions. The techniques are mainly categorized into machine learning techniques and statistical methods (Adebiyi et al, 2014). In recent years, it has been proven that machine learning techniques are capable of identifying nonlinear and volatile patterns in financial time-series (Tabachnick and Fidell, 2013). Long Short-Term Memory (LSTM) is an emerging deep learning technique.

The work we presented used the LSTM (Bengio et al., 1994) model and predicted the stock returns of the TAIEX, BSE, & KOSPI. We gathered five years of actual and predicted BSE, KOSPI &TAIEX data and utilized it to train and validate the model. The methodology portion of the paper follows, in which we will go through each phase in depth.

The rest of the paper is set out as follows: Section 7.2 is devoted to research materials; the methodology is explained in Section7.3; in Section 7.4 the experimental results and discussion are discussed; and in Section 7.5 the conclusion is formed.

One of the most powerful models for processing sequential data has been recurrent neural networks (RNN) model (Hüsken and Stagge, 2003).After providing greater prediction accuracy, RNN models have presented a serious challenge to statistical approaches for time-series forecasting issues (Zhang et al., 1998; Connor et al., 1994).

One of the most effective RNN architectures is long short-term memory. In the hidden layer of the network, LSTM (Bengio et al., 1994) adds the memory cell, a computational unit that replaces conventional artificial neurons. Networks can efficiently correlate memories and input remote in time with these memory cells, allowing them to understand the structure of data dynamically through time with great prediction ability. Two of the primary criteria of an index are that it is investable and transparent (Lo, 2016). Since financial time-series are highly volatile, noisy (Fishcher, 2018), nonlinear and dynamic in nature (Lahmiri, 2012), stock price prediction is one of the most challenging tasks, as mentioned earlier in this paper.

7.2 MATERIALS

Three datasets, TAIEX and BSE SENSEX are used for MFTS. We have taken the data from January 1 to October 31 as testing data and November1 to December 31 as testing data for each year. Where datasets, TAIEX, KOSPI and BSE SENSEX are used for MFTS.

7.3 RESEARCH METHODOLOGY

In this experiment, we have used the TAIEX, BSE & KOSPI datasets with Deep Learning-based Stacked Long Short-Term Memory (LSTM) (Bengio et al.,1994), which is an RNN architecture, for classifying, processing, and generating predictions based on time-series data. Different types of neural networks may be created by combining various parameters such as network architecture, training technique, and so on. Long Short-Term Memory was considered for this research. However, the LSTM (Bengio et al., 1994) demonstrates a type of recurrent neural network capable of learning order dependence in sequence prediction problems. Neural network focuses the highlights as single information, in any case, of the LSTM (Bengio et al., 1994) and it organizes the highlights as arrangements of information, such as discourse, video, and time-series information etc. Comparable to conventional RNNs, each LSTM (Bengio et al., 1994) arrangement is composed of an input layer, a covered-up layer(s), and a yield layer. Not at all like RNNs, the covered-up layer(s) of the LSTM (Bengio et al., 1994) are built as memory cells. Each memory cell comprises a disregard entryway, an input entryway, and a yield door. The three doors work together to direct the stream of data and to dispose of insignificant data between memory cells.

1) Disregard door (ft): Characterizes which data to be evacuated from the cell state.
2) Input gate (it): It defines which information to be added to the cell state.
3) Yield entryway (ot): It characterizes which data to be utilized as a yield.

Figure 7.1 outlines the stream of data at time t. Each entryway was considered as a neuron in a multi-layer neural organizes. Each door was related with an enactment work to compute a weighted whole. The 3 exit bolts from the memory cell to the 3 entry ways speak to the peephole associations. They indicate the commitments of the actuation of the memory cell c, at time t-1. It implies that the calculations made at step t too consider the actuation of the memory cell at time step t-1.

All the activation functions involved in equation are explained as follows:

(1) The value should be within the range of 0 and 1 to completely forget activation function scales and to completely remember, the value should be within the range of -1 and 1.
(2) σ_g: sigmoid function = $e^x / (e^x + 1)$
(3) σ_e: *hyperbolicfunction* = $\sigma_e(x) = (e^{2x} - 1) / (e^{2x} + 1)$
(4) o: Hadamard product

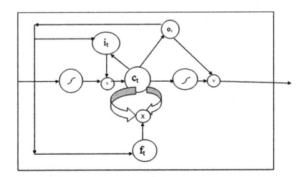

FIGURE 7.1   The architecture of the LSTM network.

At time t, firstly, the LSTM layer takes within the information from the past cell state ct-1. The actuation values ft of the forget entryways at time t are computed based on it, ht-1 and bf. Secondly, the LSTM organize decides which information should be included in the current cell state ct by computing it. In addition, ct was calculated as the sum of the element-wise item of each portion. At last, the output ht was computed. Similar to other machine learning strategies, the weight frame works Ws and the predisposition vectors bs were prepared in the iterations such that they minimized the misfortune work, Mean Squared Mistake (MSE) over the preparing tests.

7.4   RESULT AND DISCUSSION

The demonstration is executed utilizing the LSTM calculation. It is the 1-stacked layer show with root implies square blunder values. The table shows the outcomes of the datasets. The table is actualized utilizing the LSTM calculation. In this segment, this paper applies the proposed strategy to estimate the TAIEX, BSE, and KOSPI from 2015 to 2020. It presents the exploratory outcomes of the proposed strategy with all of the strategies displayed in Figures 7.2 to 7.19, showing information for actual/forecasted results over a long time frame. This paper assesses the execution of the proposed strategy utilizing RMSE, which is characterized as taking after:

$$RMSE = \sqrt{\dfrac{\sum_{i=1}^{n}(\text{Actual value}_i - \text{Forecasted value}_i)^2}{n}}$$

Where n signifies the number of days required to be forecasted. Table 7.1 shows the comparison of the RMSE and the normal RMSE for diverse strategies. The RMSE values of TAIEX 2015–2020 are shown in Figures 7.2 to 7.7, utilizing genuine information and forecasted information. Figures 7.8 to 7.13 show the RMSE values of BSE 2015–2020, utilizing genuine information and forecasted information, and Figures 7.14 to 7.19 show the RMSE values of KOSPI 2015–2020, utilizing real information and forecasted information.

Table 7.2 shows the comparison of the RMSE values with various models from the year 2015 to 2020.

TABLE 7.1
Comparisons of the RMSEs and the Average RMSE for Different Methods

Method	Year						
	2015	2016	2017	2018	2019	2020	Average
TAIEX	377.63	472.69	494.30	467.14	503.62	579.89	482.46
BSE	1150.70	1244.68	1445.15	1580.07	1660.83	1928.25	1501.61
KOSPI	96.38	93.18	120.90	89.85	94.42	125.71	103.40

TABLE 7.2
Comparison Table for TAIEX Method

Method	Year						Average RMSE
	2015	2016	2017	2018	2019	2020	
Huarng et al's (Kun-Huang Huarng et al., 2007)	627.12	505.21	708.12	607.12	805.23	789.23	673.67
Chen's Fuzzy Time Series (Shyi-Ming Chen, 1996)	467.12	985.21	832.12	537.12	572.23	689.23	679.50
Univariate Conventional Regression (Yu & Huarng, 2010)	136.30	143.18	140.90	100.85	167.42	155.71	140.72
Univariate Neural Network (Yu & Huarng, 2010)	196.38	193.18	149.50	200.85	120.42	137.71	166.34
Bivariate Conventional Regression (Yu & Huarng, 2010)	227.12	305.21	392.12	392.12	402.23	479.23	554.50
Bivariate Neural Network (Yu & Huarng, 2010)	427.12	505.21	602.12	507.12	702.23	789.23	588.83
Proposed Model	377.63	472.69	494.30	467.14	503.62	579.89	482.46

GRAPHICAL REPRESENTATION OF **TAIEX**

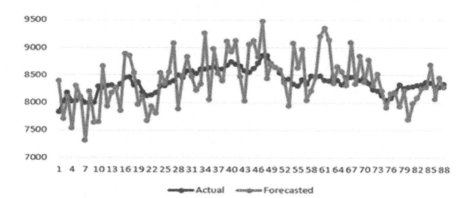

FIGURE 7.2   TAIEX forecasting graph in 2015.

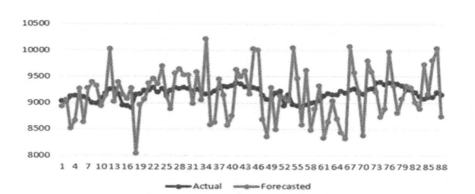

FIGURE 7.3   TAIEX forecasting graph in 2016.

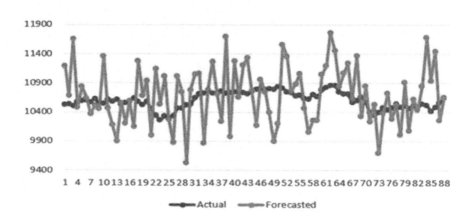

FIGURE 7.4   TAIEX forecasting graph in 2017.

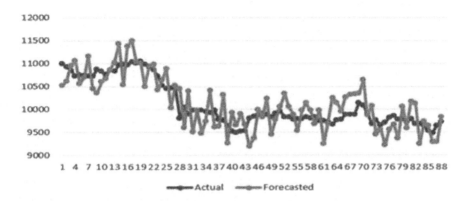

FIGURE 7.5 TAIEX forecasting graph in 2018.

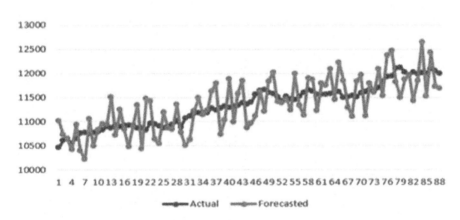

FIGURE 7.6 TAIEX forecasting graph in 2019.

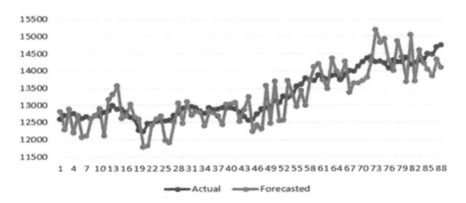

FIGURE 7.7 TAIEX forecasting graph in 2020.

TABLE 7.3
Comparison Table for BSE Method

Method	Year						Average RMSE
	2015	2016	2017	2018	2019	2020	
Huarng et al's (Kun-Huang Huarng et al., 2007)	909.35	707.24	1240.10	1222.21	1532.26	1229.20	1140.06
Chen's Fuzzy Time Series (Shyi-Ming Chen, 1996)	1188.12	1890.12	1560.15	1768.23	1990.21	2009.12	1734.32
Univariate Conventional Regression (Yu & Huarng, 2010)	1146.38	993.18	1149.50	890.85	970.42	997.71	1024.67
Univariate Neural Network (Yu & Huarng, 2010)	1736.30	1643.18	1540.90	1900.85	1647.42	1255.71	1620.72
Bivariate Conventional Regression (Yu & Huarng, 2010)	958.12	966.17	968.13	845.23	730.21	667.12	855.83
Bivariate Neural Network (Yu & Huarng, 2010)	1167.12	1134.12	1766.15	1623.23	1999.21	2059.12	1624.82
Proposed Model	1150.70	1244.68	1445.15	1580.07	1660.83	1928.25	1501.61

The comparison table (Table 7.3) shows the above average of all the methods with year 2015 to 2020 for TAIEX, which is also shown in the table inside. Figures 7.2 to 7.7 show an example of TAIEX forecasting, with data from 2015, used for the proposed method. The training dataset is from the year 2015 to 2020. In Figure 7.2 the graph indicates the actual and forecasted data using TAIEX method in 2015. In Figure 7.3 the graph indicates the actual and forecasted data using TAIEX method in 2016. In Figure 7.4 the graph indicates the actual and forecasted data using TAIEX method in 2017. In Figure 7.5 the graph indicates the actual and forecasted data using TAIEX method in 2018. In Figure 7.6 the graph indicates the actual and forecasted data using TAIEX method in 2019. In Figure 7.7 the graph indicates the actual and forecasted data using TAIEX method in 2020.

GRAPHICAL REPRESENTATION OF BSE

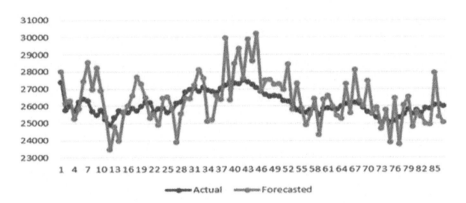

FIGURE 7.8　　BSE forecasting graph in 2015.

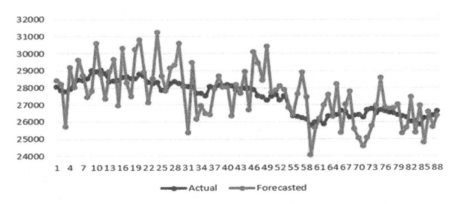

FIGURE 7.9　　BSE forecasting graph in 2016.

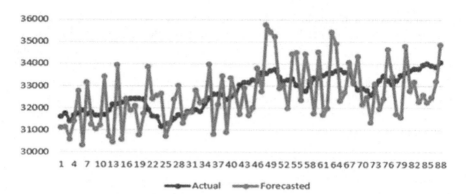

FIGURE 7.10   BSE forecasting graph in 2017.

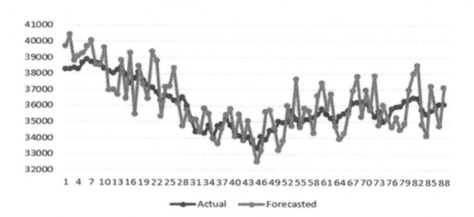

FIGURE 7.11   BSE forecasting graph in 2018.

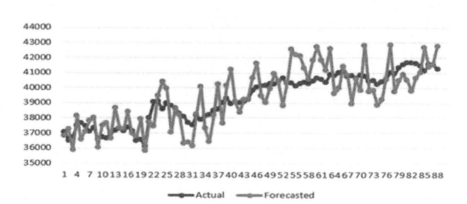

FIGURE 7.12   BSE forecasting graph in 2019.

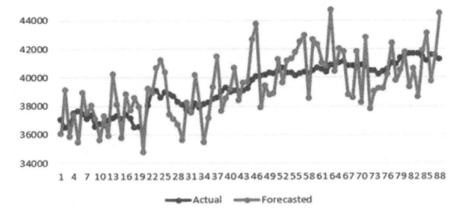

FIGURE 7.13   BSE forecasting graph in 2020.

The graphical representations show the data of BSE Limited which is also known as the Bombay Stock Exchange, which is an Indian stock exchange located on Dalal Street in Mumbai. Established in 1875 by cotton merchant Premchand Roychand, a Rajasthani Jain businessman, it is the oldest stock exchange in Asia, and also the tenth oldest in the world. The training dataset is from the year 2015 to 2020 in Figures 7.8 to 7.13. In Figure 7.8 the graph indicates the actual and forecasted data using BSE method in 2015. In Figure 7.9 the graph indicates the actual and forecasted data using BSE method in 2016. In Figure 7.10 the graph indicates the actual and forecasted data using BSE method in 2017. In Figure 7.11 the graph indicates the actual and forecasted data using BSE method in 2018. In Figure 7.12 the graph indicates the actual and forecasted data using BSE method in 2019. In Figure 7.13 the graph indicates the actual and forecasted data using BSE method in 2020.

TABLE 7.4
Comparison Table for KOSPI Method

Method	Year							Average RMSE
	2015	2016	2017	2018	2019	2020	
Huarng et al's (Kun-Huang Huarng et al., 2007)	109.33	107.21	140.12	122.22	132.27	129.21	123.39
Chen's Fuzzy Time Series (Shyi-Ming Chen, 1996)	88.12	76.12	60.15	68.23	90.21	109.12	84.46
Univariate Conventional Regression (Yu & Huarng, 2010)	196.38	193.18	149.50	200.85	120.42	137.71	166.34
Univariate Neural Network (Yu & Huarng, 2010)	136.30	143.18	140.90	100.85	167.42	155.71	140.72
Bivariate Conventional Regression (Yu & Huarng, 2010)	58.12	66.17	68.13	45.23	30.21	67.12	55.83
Bivariate Neural Network (Yu & Huarng, 2010)	167.12	134.12	66.15	23.23	99.21	159.12	108.15
Proposed Model	96.38	93.18	120.90	89.85	94.42	125.71	103.40

GRAPHICAL REPRESNTATION OF KOSPI

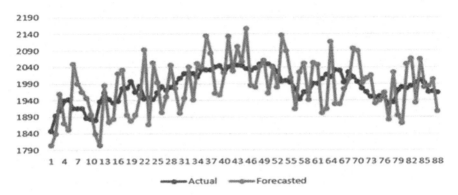

FIGURE 7.14   KOSPI forecasting graph in 2015.

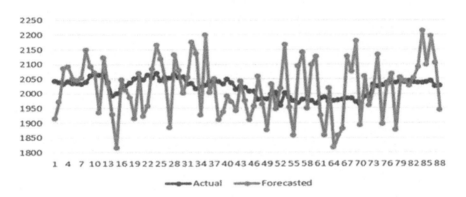

FIGURE 7.15   KOSPI forecasting graph in 2016.

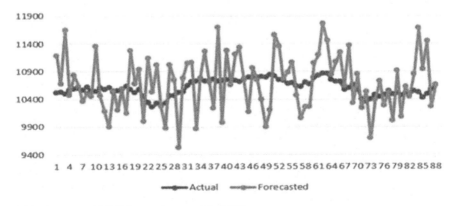

FIGURE 7.16   KOSPI forecasting graph in 2017.

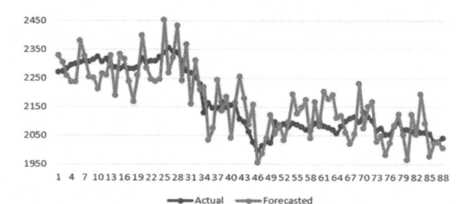

FIGURE 7.17   KOSPI forecasting graph in 2018.

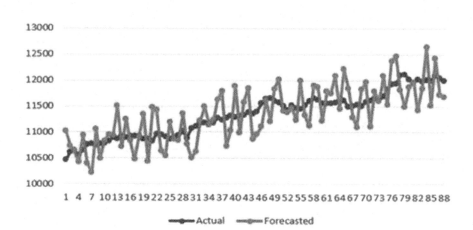

FIGURE 7.18   KOSPI forecasting graph in 2019.

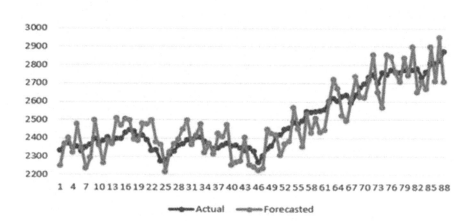

FIGURE 7.19   KOSPI forecasting graph in 2020.

The dataset from year 2015 to 2020 of KOSPI which is the Korea Composite Stock Price Index is the indexes of all common stocks traded on the Stock Market Division – previously, Korea Stock Exchange – of the Korea Exchange are being examined in this paper. It is the representative stock market index of South Korea, like the S&P 500 in the United States. The dataset from year 2015 to 2020 of KOSPI are being examined in this paper. In Figure 7.14 the graph indicates the actual and forecasted data using KOSPI method in 2015. In Figure 7.15 the graph indicates the actual and forecasted data using KOSPI method in 2016. In Figure 7.16 the graph indicates the actual and forecasted data using KOSPI method in 2017. In Figure 7.17 the graph indicates the actual and forecasted data using KOSPI method in 2018. In Figure 7.18 the graph indicates the actual and forecasted data using KOSPI method in 2019. In Figure 7.19 the graph indicates the actual and forecasted data using KOSPI method in 2020.

7.5 CONCLUSION

The LSTM model demonstrates an algorithmic approach to analyzing time arrangement information and treating the basic process. In this way the LSTM was able to make more precise expectations on stock cost movements compared to the CART demonstrate. This is often since the LSTM model, by its nature, uses a profound learning approach and is good at handling successive information, and extract useful information while dropping pointless data, and it comes with superior results and benefits. Not at all like past strategy, this approach is all about employing a factual strategy to look at the determining execution of both the proposed strategy and other strategies. Beginning with the case of TAIEX estimating, the proposed strategy is found to offer superior determining execution than the other strategies displayed since 2017 and there appears to be no contrast after 2017. Besides, no factual comparison is found between the strategies created after 2017 and the proposed strategy. One of the reasons is that the proposed strategy as it were, employed one calculation to estimate the TAIEX, whereas other strategies utilize one or more variables, which may move forward the determining execution. Although the proposed strategy as it were, employed one figure, a conventional procedure for dividing the approach, and second-order record the information to estimate the stock file and from the estimating comes about of this paper, utilizing three databases of TAIEX, BSE, and KOSPI, the proposed strategy is found to offer superior estimating execution. To construct the numerous straight models of the dataset, a novel high-order time arrangement these model calculation is utilized to gather information by their direct connections rather than shapes, which builds a more appropriate direct demonstration than other group-based models. The proposed model calculation is compared with the other three model calculations and built the finest reasonable direct demonstrate which gives the accurate result. From a workable point of view, for speculators interested in choosing between measurable strategies, or machine learning techniques, or profound learning procedures for determining which stocks to purchase, we propose speculators utilize the LSTM model for their stock cost forecasts because of the deep learning approach of the LSTM show. Finally, this paper offers a determining show based on ANN, which calculates the weight of related numerous straight models, as well as its learning calculation. The proposed estimating show is compared with another FTS-based demonstration on the TAIEX, BSE, KOSPI and it gave a much higher, stronger and improved determining precision than others. Too, by comparing with estimating models that are FTS-based, the outcome shows that determining demonstrate seems to handle the inadequate, uncertain information and shows an improved robustness. The RMSE esteem of TAIEX is 482.46, BSE is 1501.61& KOSPI is 103.40. It is proposed that further research be performed to better understand the execution of the LSTM demonstrated in other places of market.

REFERENCES

Adebiyi, Ayodele Ariyo; Adewumi, Aderemi Oluyinka; Ayo, Charles Korede (2014). Comparison of ARIMA and Artificial Neural Networks Models for Stock Price Prediction. *Journal of Applied Mathematics*, 2014(), 1–7. doi:10.1155/2014/614342.

Barak, Sasan; Modarres, Mohammad (2015). *Developing an approach to evaluate stocks by forecasting effective features with data mining methods. Expert Systems with Applications*, 42(3), 1325–1339. doi:10.1016/j.eswa.2014.09.026.

Bengio, Y.; Simard, P.; Frasconi, P. (1994). Learning long-term dependencies with gradient descent is difficult. *IEEE Transactions on Neural Networks*, 5(2), 0–166. doi:10.1109/72.279181.

Chen, Shyi-Ming (1996). Forecasting enrollments based on fuzzy time series. *Fuzzy Set and Systems*, 81(3), 311–319. doi:10.1016/0165–0114(95)00220–0.

Connor, J. T., Martin, R. D., & Atlas, L. E. (1994). *Recurrent neural networks and robust time series prediction. IEEE Transactions on Neural Networks*, 5(2), 240–254. doi:10.1109/72.279188.

Fischer, T., Krauss, C (2018). Deep learning with long-shortterm-memory networks for financial market predictions. *European Journal of Operational Research*, 270(2), 654–669. 10.1016/j.ejor.2017.11.054.

Huarng, Kun-Huang, Tiffany Hui-Kuang Yu, & Yu Wei Hsu (2007). A Multivariate Heuristic Model for Fuzzy Time-Series Forecasting. *IEEE Transactions on Systems, Man, and Cybernetics, Part B (Cybernetics)*, 37(4), 0–846. doi:10.1109/tsmcb.2006.890303.

Hüsken, M. & P. Stagge, (2003) Recurrent neural networks for time series classification, *Neurocomputing*, 50(C), 223–235.

Kim, Kyoung-jae & Ingoo Han (2000). Genetic algorithms approach to feature discretization in artificial neural networks for the prediction of stock price index. *Expert Systems with Applications*, 19(2), 125–132. doi:10.1016/s0957–4174(00)00027–0.

Lahmiri, S (2012). Linear and nonlinear dynamic systems in financial time series prediction. *Management Science Letters*, 2551–2556.

Lo, Andrew W. (2016). What Is an Index? *The Journal of Portfolio Management*, 42(2), 21–36. doi:10.3905/jpm.2016.42.2.021.

Tabachnick, B. G., & Fidell, L. S (2013). *Using Multivariate Statistics*. Upper Saddle River, NJ, USA, 6th edition.

Yu, T. H. K. & Huarng, K. H. (2010). Corrigendum to "A bivariate fuzzy time series model to forecast the TAIEX." *Expert Systems with Applications* 34(4), 2945–2952. DOI: 10.1007/978–3–642–38577–3_48.

Zhang, Guoqiang., B. Eddy Patuwo; Michael Y. Hu (1998). Forecasting with artificial neural networks: The state of the art. *International Journal of Forecasting* 14(1), 35–62. doi:10.1016/s0169–2070(97)00044–7.

8 A Comparative Study and Analysis of Time-Series and Deep Learning Algorithms for Bitcoin Price Prediction

*Sagar Chakraborty
Seacom Engineering College, Kolkata, India

Nivedita Das
Techno India Group, Kolkata, India

Kisor Ray
Techno India University, Kolkata, India

*Corresponding author.

CONTENTS

DOI: 10.1201/9781003279044-8

8.1   INTRODUCTION

Bitcoin (Nakamoto, 2008) is the first and the leading cryptocurrency which was created by an anonymous person or group named Satoshi Nakamoto in the year 2008 using the blockchain platform. It has a current market capitalization of 890 billion USD according to CoinMarketCap (https://coinmarketcap.com/). Companies like Microstrategy, Tesla, Galaxy Digital Holdings, etc. are the major investors in Bitcoin. Bitcoin price was close to 0 USD when it was launched in the year 2009. However, the current Bitcoin price is almost 48000 USD. Bitcoin is a decentralized cryptocurrency that is used world-wide for digital payments or investments. Traders can invest using various marketplaces which are commonly known as "Bitcoin Exchanges." These exchanges offer traders to sell or buy Bitcoins using other currencies. At present Binance is the largest Bitcoin exchange. All the transaction records and the timestamp data are stored in a constantly growing ledger which is known as Blockchain. Each block represents a record in a blockchain. Each block contains a pointer that is used to point to the previous block of data. The data on the blockchain is secure and immutable. During any transaction, the trader's identity is not revealed, but only the wallet ID is made public.

Bitcoin provides an opportunity for price prediction due to its relatively early stage and price volatility, which is far much higher than that of traditional fiat currencies. A number of researches have been made for the full-grown financial markets such as the stock market. Typically, traditional time-series prediction methods depend on trends, seasonal variation, and noise. This methodology is more appropriate for forecasting data where seasonal effects are present. But due to the lack of seasonality and high volatility in the Bitcoin market, these methods are not very effective. Bitcoin can be considered a financial asset and it can be traded using other cryptocurrency exchanges.

For the last few years, researchers have been trying to investigate the aspects that directly or indirectly affect the Bitcoin price and the shapes behind its volatility using various algorithms and analytical tools. Due to the recent advances in machine learning and deep learning for forecasting, several Bitcoin predictions models have been proposed by researchers. Although many machine learning algorithms were used for the Bitcoin price prediction, most previous work considered only a few deep learning methods such as deep neural network (DNN) or recurrent neural network (RNN).

In this paper, we study and compare the time-series ARIMA model, linear regression, LASSO regression, and various futuristic deep learning methods, such as RNNs, long short-term memory (LSTM), and their integrated models for Bitcoin price prediction. We developed various regression models by exploiting the Bitcoin historical data and compared their prediction performance considering daily and monthly data. Experimental results showed that LASSO and LSTM prediction models marginally surpassed the other prediction models for regression problems. Bitcoin daily closing price in USD from 17 October 2017 to 12 November 2021 is shown in Figure 8.1.

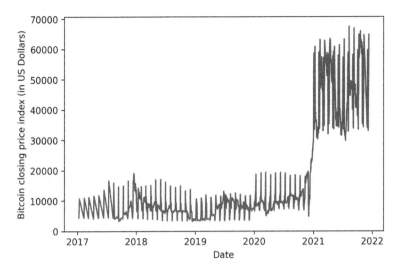

FIGURE 8.1   Bitcoin daily closing price in USD from17 October 2017 to 12 November 2021.

8.2   LITERATURE SURVEY

8.2.1   RELATED WORKS

Research on Bitcoin is relatively low, though it has gained popularity in the last few years. The literature survey covers work done on Bitcoin price prediction using various machine learning algorithms. Sean McNally, J. Roche and S. Caton (2018), in their paper "Predicting the Price of Bitcoin Using Machine Learning", employed RNN, LSTM, and ARIMA model and it is found that LSTM achieved the highest accuracy while the RNN achieved the lowest RMSE but ARIMA performed poorly in terms of accuracy and RMSE. Bruno Miranda Henrique, Vinicius Amorim Sobreiro and Herbert Kimura (2019) in their manuscript "Literature review: Machine learning techniques applied to financial market prediction," proposes the use of bibliographic survey techniques that are applied to the literature about machine learning for predicting financial market values. Huisu Jang and Jaewook Lee (2017) in their study, "An empirical study on modeling and prediction of Bitcoin prices with Bayesian Neural Networks (BNN) based on Blockchain information," unveils the effect of Bayesian neural networks by evaluating and analyzing the time-series of the Bitcoin process selecting most appropriate features from Blockchain information. Yiqing Hua (2020), in his research "Bitcoin price prediction using ARIMA and LSTM" trained ARIMA and LSTM models with real-time price data and predicted the accuracy of the models. Lekkala Sreekanth Reddy and P. Sriramya (2020), in their paper "Bitcoin Price Prediction Using LASSO Algorithm." has applied the LASSO regression model to predict the Bitcoin price. In the study, he found that the accuracy of the model is 97%. Siddhi Velankar, Sakshi Valecha, and Shreya Maji (2018), in their study "Bitcoin price prediction using machine learning", have applied Bayesian Regression and Generalized Linear Model (GLM) to find the Bitcoin price accurately. Radityo, Munajat, and Budi (2017), in their research "Prediction of Bitcoin exchange rate to American dollar using artificial neural network methods," implemented BPNN, GANN, GABPNN, NEAT methods on the historical data and found BPNN is the best performing method. Brownlee (2016), in his study "Time series prediction with LSTM Recurrent Neural Networks in Python with Keras" implemented the LSTM model to predict the Bitcoin price.

8.2.2 BLOCKCHAIN TECHNOLOGY

Though blockchain (Nofer et al., 2017; Zheng et al., 2017) technology was introduced by the researchers Stuart Haber and W. Scott Stornetta in the year 1991, it was first applied in the year 2009 by Satoshi Nakamoto when Bitcoin was launched. The blockchain is the primary building block for Bitcoin. Blockchain is a decentralized continuously growing ledger that chains all the mined blocks in chronological order. This chain is secure and unchangeable. A sample blockchain formation is shown in Figure 8.2.

8.3 METHODOLOGY

The proposed methodology considers the time-series ARIMA model, LASSO regression, and LSTM models to forecast the daily price of Bitcoin by identifying and evaluating relevant features of the dataset. After analyzing the performance of all the Bitcoin prediction models, we can determine which model is much more accurate for the future fulfillment of our target and select appropriate parameters to obtain a better performance. The workflow of the Bitcoin prediction using machine learning models is shown in Figure 8.3.

8.3.1 MACHINE LEARNING MODELS

8.3.1.1 ARIMA Model
ARIMA (Autoregressive Integrated Moving Average) model is the integration of two models – Autoregressive (AR) and Moving Average (MA). ARIMA model is defined by three terms -p, d, and q where "p" is the order of the "Auto Regressive" (AR) term, "d" is the minimum number of differences needed to make the series stationary, and "q" is the order of the "Moving Average" (MA) term.

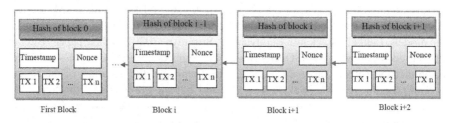

FIGURE 8.2 Sample blockchain formation.

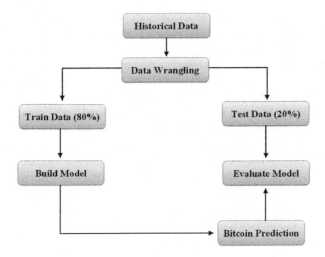

FIGURE 8.3   Workflow of the Bitcoin prediction using machine learning models.

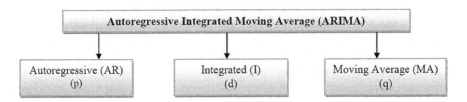

FIGURE 8.4   ARIMA time-series model.

Autoregressive (AR) model is a time-series model where earlier values are considered as input into a regression equation for predicting the current values.

$$y_t = \beta_1 y_{t-1} + \beta_2 y_{t-2} + \beta_3 y_{t-3} + \ldots + \beta_p y_{t-p} \tag{1}$$

Moving Average (MA) model is a time-series linear regression of the current value against present and the previous error terms.

$$y_t = \varepsilon_t + \alpha_1 \varepsilon_{t-1} + \alpha_2 \varepsilon_{t-2} + \ldots + \alpha_q \varepsilon_{t-q} \tag{2}$$

In this study, we use both non-seasonal ARIMA and seasonal ARIMA(SARIMA) models for forecasting Bitcoin prices. Here we have used the auto_arima function to detect the parameters automatically.

8.3.1.2   LASSO Regression

The LASSO (Least Absolute Shrinkage Selection Operator) regression model can be used to predict the Bitcoin price. In the LASSO model, a small amount of bias is introduced to minimize the variance. Bias is the inability of a machine learning algorithm (like linear regression) to capture the true relationship between independent and dependent variables. The difference in fits between training and testing datasets is called variance. Compared to straight-line squiggly line fits greatly to the training set, resulting in MSE = 0. But when MSE calculated with respect to testing set, straight line (linear regression) performs better. The squiggly line has a low bias since it is flexible and can be adapted to the curve in the relationship between the independent and dependent variables. But, it has high variability because it results in a vastly different sum of squares for different datasets. So, it is hard to predict how well the squiggly line will perform with future datasets. Sometimes its performance is very poor. In contrast, the straight line has a relatively high bias, since it cannot capture the curve in the relationship between the independent and dependent variables, but it has a relatively low variance because the sum of squares is very similar for different datasets. The straight line might only give good predictions but not great predictions. In contrast, the straight line has a relatively high bias, since it cannot capture the curve in the relationship between the independent and dependent variables, but it has a relatively low variance because the sum of squares is very similar for different datasets. The straight line might only give good predictions but not great predictions. In machine learning, the ideal algorithm must have low bias and can accurately model the true relationship and it has low variability, by producing consistent predictions across different datasets. This can be achieved by the LASSO algorithm which is used to find the sweet spot between a simple model and a complex model using the following methods -regularization, boosting, and bagging.

8.3.1.3   Recurrent Neural Network (RNN)

RNN (recurrent neural network) is the class of Artificial Neural Network (ANN) which is mainly used for natural language processing (NLP) and sentiment analysis. RNN overcomes the problems that are faced by Feed Forward Neural Networks, not designed for sequences/time-series data and does not model memory. RNN works on sequence data and its performance in time-series prediction is very good. Recurrent Neural Network (RNN) works on the following recursive function $S_t = F_w (S_t, X_t)$, where X_t is the input at time step t, S_t is the State at time step t and F_w is Recursive function. The workflow of a simple RNN model is shown in Figure 8.5.

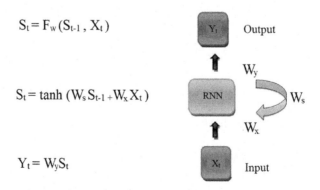

$$S_t = F_w (S_{t-1}, X_t)$$

$$S_t = \tanh (W_s S_{t-1} + W_x X_t)$$

$$Y_t = W_y S_t$$

FIGURE 8.5   Workflow of simple RNN model.

8.3.1.4 Long Short-Term Memory (LSTM)

LSTM (long short-term memory) is a special type of recurrent neural network that is designed to overcome the limitations of the RNN model:

(i)   Gradient vanishing and exploding
(ii) Complex training
(iii) Difficulty to process very long sequences

The working principle of a simple RNN model and LSTM model is shown in Figure 8.6 and Figure 8.7, respectively.

LSTM uses both the sigmoid (σ) and tanh functions. The sigmoid function is used to forget and remember information. Sigmoid gives the output from 0 to 1. The tanh function is used to overcome the vanishing gradient problem. The tanh's second derivative can sustain for a long-range before going to zero.

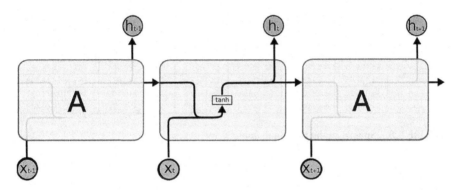

FIGURE 8.6 Working principle of simple RNN model.

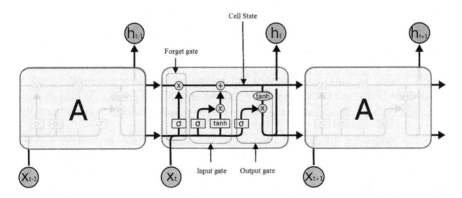

FIGURE 8.7 Working principle of LSTM model.

TABLE 8.1
Features of the Bitcoin Dataset

Features	Description
Date	Date
Open	Open price at start time window
High	High price within time window
Low	Low price within time window
Close	Close price at end of time window
Volume	Volume of BTC transacted in this window
Market cap	Total Market Capitalization

Date	Open*	High	Low	Close**	Volume	Market Cap
27-Aug-17	$4,345.10	$4,416.59	$4,317.29	$4,382.88	$1,537,459,968	$72,441,993,792
26-Aug-17	$4,372.06	$4,379.28	$4,269.52	$4,352.40	$1,511,609,984	$71,929,882,019
25-Aug-17	$4,332.82	$4,455.70	$4,307.35	$4,371.60	$1,727,970,048	$72,241,507,489
24-Aug-17	$4,137.60	$4,376.39	$4,130.26	$4,334.68	$2,037,750,016	$71,625,870,868
23-Aug-17	$4,089.01	$4,255.78	$4,078.41	$4,151.52	$2,369,819,904	$68,593,954,170
22-Aug-17	$3,998.35	$4,128.76	$3,674.58	$4,100.52	$3,764,239,872	$67,743,202,431
21-Aug-17	$4,090.48	$4,109.14	$3,988.60	$4,001.74	$2,800,890,112	$66,106,992,038
20-Aug-17	$4,189.31	$4,196.29	$4,069.88	$4,087.66	$2,109,769,984	$67,520,478,838
19-Aug-17	$4,137.75	$4,243.26	$3,970.55	$4,193.70	$2,975,820,032	$69,265,141,208
18-Aug-17	$4,324.34	$4,370.13	$4,015.40	$4,160.62	$2,941,710,080	$68,710,871,037
17-Aug-17	$4,384.44	$4,484.70	$4,243.71	$4,331.69	$2,553,359,872	$71,527,949,443

FIGURE 8.8   Snippet of the Bitcoin dataset.

8.3.2 Dataset

In this study, we have collected the Bitcoin dataset from https://coinmarketcap.com for 1548 days (17 August 2017 to 12 November 2021). The features that are used for Bitcoin price prediction are shown in Table 8.1.

A snippet of the Bitcoin Dataset is shown in the Figure 8.8.

8.4   RESULTS AND DISCUSSIONS

In this study, we have implemented the following forecasting models- ARIMA, SARIMA, LSTM, and LASSO on daily and monthly data to predict the bitcoin price. The entire coding part is done in the Python language in the Google Colaboratory platform. For measuring the performance, we used R-squared error $\left(R^2\right) = 1 - \dfrac{SS_{Regression}}{SS_{Total}}$

where $SS_{Regression}$ is the sum squared regression error and SS_{Total} is the Sum Squared Total Error. In this study, we have done univariate time-series analysis by using the "close" price in our dataset but we have an aim to include some more features in the future to improve the forecast. However, for the LASSO regression model, we have considered three features, "high," "low," and "open" price of bitcoin to predict the "close" price. Bitcoin closing price plot on per day, per month, per quarter and per year basis is shown in Figure 8.9.

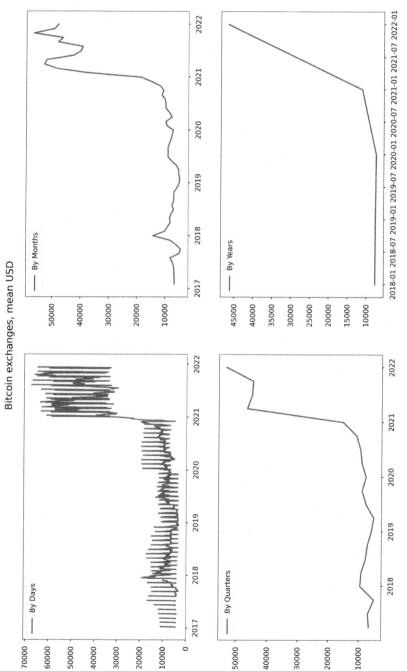

FIGURE 8.9   Bitcoin closing price plot on per day, month, quarter and year basis.

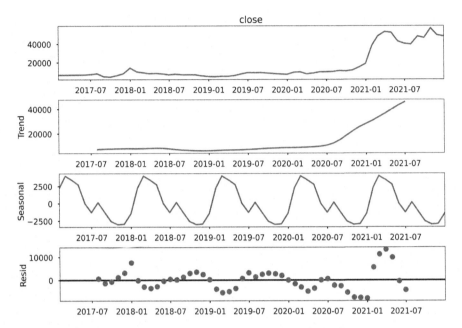

FIGURE 8.10   Plot for identification of seasonality and trend in the dataset.

Then we have performed seasonal decomposition on monthly data to find if there exists any seasonality or repeating pattern in our Bitcoin dataset.

From the seasonal graph in Figure 8.10 we can clearly see some seasonality is present in the monthly data and also there is an upward trend in the data that we can find from the trend graph.

8.4.1   ARIMA – TIME-SERIES MODEL

On Daily Data

To implement the ARIMA model we used numpy, pandas, matplotlib, statsmodels, pmdarima, and sklearn libraries. To predict using the ARIMA model our dataset needs to be stationary (constant mean, constant variance, and no seasonality). So, we have investigated if the model is stationary or not using the Dickey-Fuller test on our daily data. As the test result P = 0.98907, which is much greater than 0.05 we can conclude that our dataset is not stationary Hence calculated the First Difference (d = 1) of our dataset to make our data stationary. Plot of stationary dataset after first difference calculation is shown in Figure 8.11.

This is the plotted form of the data after the first difference. From visual observation we can see there is a constant mean, which depicts the series now stationary but for theoretical confirmation, we again perform the Dickey-Fuller test. This time the result has given a very low value of p, depicting that the data is now stationary.

Now we have plotted ACF and PACF for daily data after the first difference in Figure 8.12.

Then we have divided the total dataset into training and testing datasets. After that, we trained the model using ARIMA and then tested the performance with the test

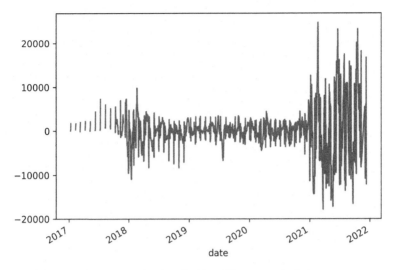

FIGURE 8.11   Plot of stationary dataset after first difference calculation.

data. ARIMA model shows us very poor performance while making predictions for the next 5 days, with an accuracy of 33%. Result of ARIMA model on daily test data is shown in Figure 8.13.

On Monthly Average Data

We have also performed a Dickey-Fuller test to find out whether the monthly data is stationary or not before applying ARIMA but the result depicted a larger p-value, which says that the model is not stationary. So, we have to make the difference 3 times to make the dataset stationary. Plot of stationary data after third difference calculation is shown in Figure 8.14 and ACF and PACF plots after third difference calculation is shown in Figure 8.15.

After that, we have implemented ARIMA on monthly average training data is shown in Figure 8.16 and tested with test data. This model also gave us a poor result too.

8.4.2   SARIMA

On Daily Data

As we have seen seasonality in our dataset so we have implemented SARIMA by introducing the seasonal component i.e., SARIMA (p,d,q) $(P,D,Q)_m$ using the training dataset. Here we have used the same packages as ARIMA. We predicted 5 days "close" values using SARIMA and when compared predicted values with the testing data, this model gives a poor performance with an accuracy of 39%. Result of SARIMA model on daily test data is shown in Figure 8.17.

On Monthly Data

Introducing the seasonal component i.e SARIMA (p,d,q) $(P,D,Q)_m$ into our monthly average training data and testing the predicted values we got R-squared Error with 43% accuracy. Result of SARIMA model on monthly average data is shown in Figure 8.18.

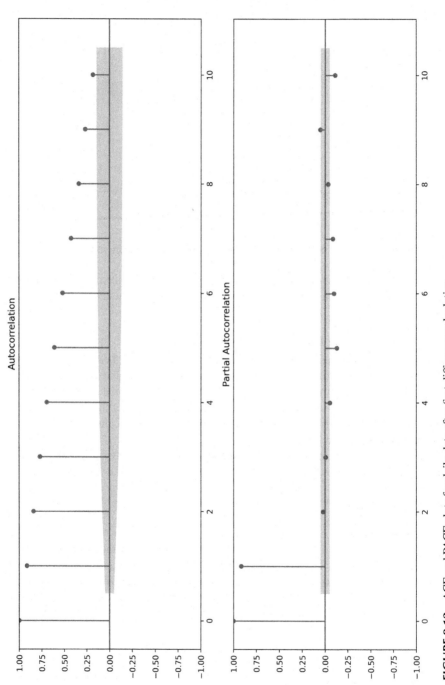

FIGURE 8.12   ACF and PACF plots for daily data after first difference calculation.

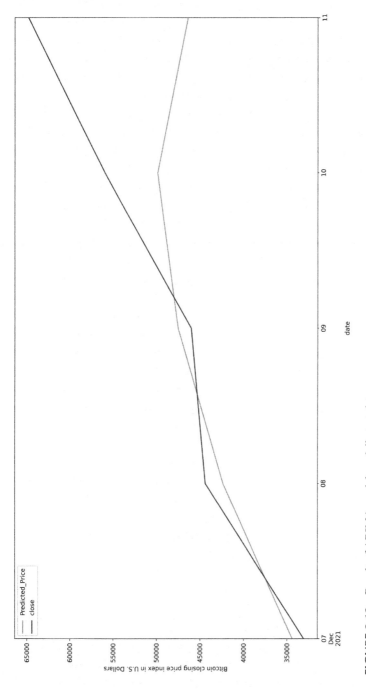

FIGURE 8.13 Result of ARIMA model on daily test data.

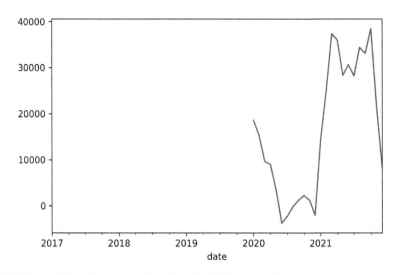

FIGURE 8.14   Plot of stationary data after third difference calculation.

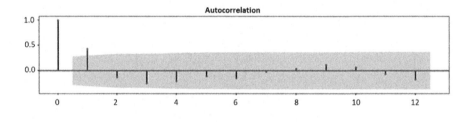

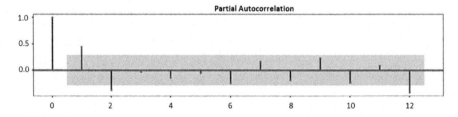

FIGURE 8.15   ACF and PACF plots after third difference calculation.

8.4.3   LSTM

For prediction using LSTM, we used numpy, pandas, matplotlib, sklearn and keras libraries. We have performed stacked LSTM on daily "close" values. Our total dataset has 1550 rows. We have divided the total dataset into two parts: training and testing. After that, we have pre-processed the data using Standard Scaler to convert the data in a scale of 0 to 1 to get accurate predictions. We used keras library to implement LSTM. When we have taken the previous 2 days "close" values to get the next day's prediction we found that accuracy with r square error function was 52% but when using 1-day

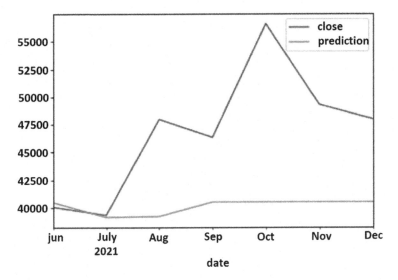

FIGURE 8.16   Result of ARIMA model on monthly average data.

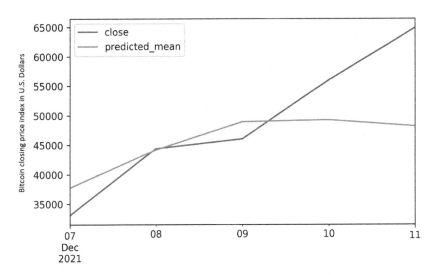

FIGURE 8.17   Result of SARIMA model on daily test data.

prior value for prediction of the next day we got 68% accuracy with 200 epochs. But when using 1-day prior value for prediction of the next day we got 70% accuracy with 700 epochs. We have used our model to predict the next 5 days "close" price.

While considering the previous 1-day "close" value as input and number of epochs equal to 700 we got the following result. Result of LSTM model on daily data is shown in Figure 8.19.

Thus, we can say that LSTM performs well on a per day "close" value.

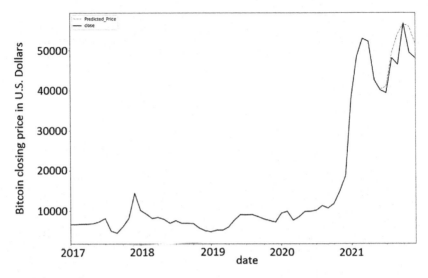

FIGURE 8.18   Result of SARIMA model on monthly average data.

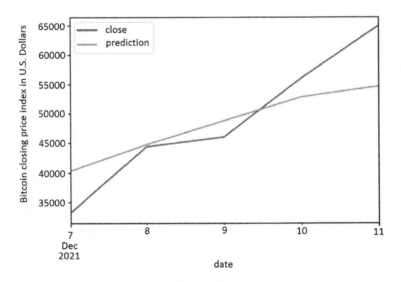

FIGURE 8.19   Result of LSTM model on daily data.

8.4.4   LASSO REGRESSION

For prediction using the LASSO regression model, we used numpy, pandas, matplotlib and sklearn libraries. We used GridSearchCV to get the optimal parameters. In LASSO Regression we have considered three features of the dataset, "high," "low" and "open" price of bitcoin for prediction of Bitcoin price and have analyzed that

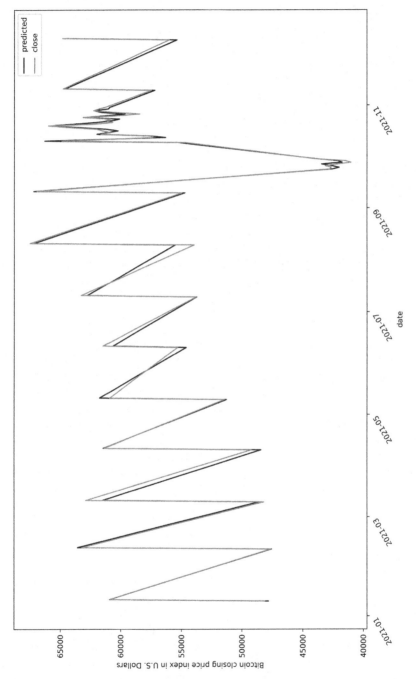

FIGURE 8.20 Result of LASSO regression model on daily data.

TABLE 8.2
R-squared Error of Each Model

Algorithm	ARIMA	SARIMA	LSTM	LASSO
R-squared error	0.33	0.39	0.70	0.99

"close" value is highly correlated with "high." We have divided our total dataset into two parts, training and testing dataset. After the train test split, we scaled the data for better performance. After training our model, we tested the predicted set of values with test data to calculate the accuracy.

We can see from Figure 8.20 that LASSO regression has performed quite well and while testing the predicted result with the actual data we got R-squared Error with 99% of accuracy.

After calculating the R-squared error of all the implemented models, we found the following result which is shown in Table 8.2.

After comparing the R-squared error of all the implemented models, we can conclude that LASSO regression and LSTM performed better compared to ARIMA, SARIMA with a high accuracy value.

8.5   FUTURE WORK

In this study, we have used historical data to predict the Bitcoin price but there are many other features that can affect the Bitcoin price like fear and greed index, panic sales, purchasing spree, government policies and regulations on cryptocurrencies, human sentiments, etc. If we train our models with the above-mentioned features along with the historical data, these models will perform better. Also, we can upgrade our models by further hyperparameter optimization. Here we have used time-series models, regression models, and some machine learning and deep learning models to predict the Bitcoin price. In the future, we can use some advanced deep learning algorithms and also hybrid deep learning models to predict the Bitcoin price. We can also implement an integrated LASSO and LSTM-based model to predict the Bitcoin price for better accuracy as both the algorithms performed better in the study.

8.6   CONCLUSION

As cryptocurrency is relatively new in the financial market and gained attention from investors in the last few years, it has lots of opportunities for researchers to apply various machine learning and deep learning models to predict Bitcoin price and analyze the performance of the models. We have applied various time-series, machine learning and deep learning models to predict the Bitcoin price considering the most relevant features from historical data. Then we performed a comparative study and analyzed the performance of each model using MSE (Mean Square Error), R Square Errors. Finally, we can say that LSTM and LASSO models perform better than other regression and time-series models. Though in our study, above-mentioned models performed better on our selected dataset, it is possible these models may fail to perform well if we consider the dataset from different ranges as Bitcoin price is fluctuating at every moment and fluctuations can be very high.

Conflict of Interests: The authors declare no conflicts of interest.

REFERENCES

Brownlee, J. 2016. "Time series prediction with LSTM Recurrent Neural Networks in Python with Keras," Available at Mach. com, p. 18.

Henrique Bruno Miranda, Vinicius Amorim Sobreiro and Herbert Kimura. 2019. "Literature review: Machine learning techniques applied to financial market prediction," *Expert Systems with Applications*, 124, pp. 226–251.

Hua, Yiqing. 2020. "Bitcoin price prediction using ARIMA and LSTM," E3S Web of Conferences 218, 01050.

Jang Huisu and Jaewook Lee. 2017. "An Empirical Study on Modeling and Prediction of Bitcoin Prices with Bayesian Neural Networks based on Blockchain Information," in *IEEE Early Access Articles*, 99, pp. 1–11.

Lekkala Sreekanth Reddy and P. Sriramya. 2020. "Bitcoin Price Prediction Using LASSO Algorithm," *Journal of Critical Reviews* 7, pp. 1188–1193.

McNally, S., J. Roche and S. Caton. 2018. "Predicting the Price of Bitcoin Using Machine Learning," 26th Euromicro International Conference on Parallel, Distributed and Network-based Processing (PDP), ambridge, pp. 339–343.

Nakamoto, S. 2008. "Bitcoin: A Peer-to-Peer Electronic Cash System. Technical Report," Satoshi Nakamoto Institute.

Nofer, Michael, Peter Gomber, Oliver Hinz and Dirk Schiereck. 2017. "Blockchain," *Business & Information Systems Engineering*, 59(3), 183–187.

Radityo, A., Q. Munajat, and I. Budi. 2017. "Prediction of Bitcoin exchange rate to American dollar using artificial neural network methods," in *Advanced Computer Science and Information Systems (ICACSIS), 2017 International Conference on*, pp. 433–438.

Velankar, Siddhi, Sakshi Valecha, and Shreya Maji. 2018. "Bitcoin price prediction using machine learning," *2018 20th International Conference on Advanced Communication Technology (ICACT)*, pp. 144–147. DOI: 10.23919/ICACT.2018.8323676.

Zheng, Z., S. Xie, H. Dai, X. Chen and H. Wang. 2017. "An Overview of Blockchain Technology: Architecture, Consensus, and Future Trends," *2017 IEEE International Congress on Big Data (BigData Congress)*, pp. 557–564.

9 Machine Learning for Healthcare

*Udeechee

School of Computer and Systems Sciences,
Jawaharlal Nehru University, New Delhi, India

T.V. Vijay Kumar

School of Computer and Systems Sciences,
Jawaharlal Nehru University, New Delhi, India

*Corresponding author.

CONTENTS

9.1   INTRODUCTION

The healthcare sector has always had a significant role in the well-being of society. However, varying availability of healthcare services has often led to unnecessary expenditure on extending such services in places where these services are not readily available. In earlier times, textbooks and experts were the only medium for accessing knowledge of the medical field, and practical knowledge was gained mainly while treating the patients. But with time and rapid technological advancements, these tasks have become simpler (Tekkeşin, 2019). Improper use of available healthcare data has led to mistakes in diagnosis and treatment resulting in an avoidable trust deficit in the existing technologies currently available with the healthcare sector (Ginneken et al., 2015).

Technology helps in analyzing data in order to reveal useful information and results. Artificial Intelligence (AI) has improved the ability as well as performance vis-à-vis decision making in various fields and subfields, and healthcare is one among them (Yu et al., 2018). The combination of experienced clinicians, readily available empirical data, and the power of AI has enhanced the capability to provide high quality healthcare services at efficient cost. Due to rapid advancements in the healthcare sector, a large amount of data is generated from different sources pertaining to genome sequencing, electronic health records, physiological matrices produced by biosensors and high-resolution medical imaging, etc. It is difficult for healthcare workers to manually analyze such large amounts of data in the limited available time and this has led to dependency on machines and technology (Ginneken et al., 2015). Further, image-based data related to radiology, ophthalmology, dermatology and pathology, generate high resolution images which, when analyzed by inexperienced clinicians, may result in incorrect diagnosis and treatments. To overcome such situations, AI-technologies can provide readily available, useful platforms to clinicians to arrive at cogent and correct diagnosis and treatment by learning from such healthcare data (Ginneken et al., 2015).

AI has various applications in the healthcare sector. These techniques are dependent on the available healthcare data. Due to advancements in technologies, a vast amount of healthcare data is available that can be utilized to extract hidden information. For example, wearable devices provide biomedical parameters related to heart rate, sleep cycle, breathing, physical activity, voice, etc., which can then be used for detecting the health conditions of individuals. Such sensors are capable of monitoring and detecting symptoms of cardiovascular disease, sleep disorder, Parkinson's, etc. (Yu et al., 2018).

Several healthcare data repositories are available, such as the World Health Organization (WHO), the Global Research and Publications Database, Kaggle, the UCI repository, Google Dataset. There is a large amount of data available in these repositories, related to different health issues and diseases such as diabetes, breast cancer, heart failure prediction, ECG heartbeat categorization, liver patients, national health and nutrition examination survey, chronic illness, COVID-19 clinical trials, diabetic retinopathy, HIV AIDS, stroke prediction, cardiovascular disease, hepatitis, drug classification, sleep, mental illness, EEG, cervical cancer, thyroid disease, etc.

Machine Learning (ML), a branch of AI, is the fastest growing field of computer science that has vast applications in various domains. It is used primarily for detection of some informative patterns hidden in data (Osisanwo et al., 2017). In recent times, ML is emerging as one of the most widely used AI techniques in the healthcare domain. It has immense potential to provide improved services at lower cost to patients and to become a helping aid to healthcare providers (Bhardwaj et al., 2017). It has the ability to predict healthcare data results with greater precision using statistical techniques and advanced algorithms. ML is widely used for analyzing large amounts of data generated in the healthcare domain. In addition, ML is also used by physicians for making accurate decisions in different subdomains of healthcare (Dhillon & Singh, 2019).

ML is a mechanism for enabling machines to learn, without being explicitly programmed (Firdaus & Hassan, 2020). It mainly uses statistical techniques and advanced algorithms. The type of algorithm used for tackling a specific problem depends upon the features in the available data and the objective to be achieved (Mahesh, 2020). ML can be classified into the following three categories namely supervised, unsupervised and reinforcement learning as shown in Figure 9.1.

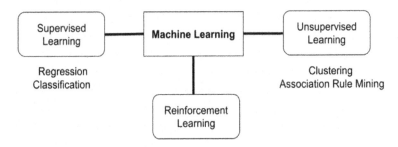

FIGURE 9.1   Types of machine learning.

In Supervised learning, algorithms are trained on labelled data for classification of new data instances (Mahesh, 2020). Supervised learning can be used for approaching the regression or classification problems. In regression models, the input space can be mapped into a continuous valued domain, whereas in classification models, it can be mapped into pre-defined classes or labels (Nasteski, 2017). Classification-based ML models are built and evaluated using two steps: training and testing. Accordingly, the dataset is split into training data and testing data. In the training phase, the inputs are taken from the training dataset in which the learning algorithms learn from the features contained within the data based on which a trained ML model is constructed. While in the testing phase, the performance of the trained ML model is evaluated using the testing data (Nasteski, 2017). Some of the most widely used supervised learning techniques include Linear Regression (R), Logistic Regression (LR), Support Vector Machine (SVM), Decision Tree (DT), Artificial Neural Network (ANN), Naïve Bayes (NB) and Random Forest (RF).

In Unsupervised learning, the target values are not known. It is called "unsupervised" as there is no target class using which the model can be trained (Mahesh, 2020). Unsupervised learning algorithms can discover hidden patterns within the data. These group instances are based on similarity and/or distance measure. The groups so created are called clusters where the clusters are formed on the basis of training data (Nasteski, 2017). Some of the most widely used unsupervised learning techniques include K-Means, DBSCAN, agglomerative and divisive clustering, Apriori, FP-Growth, Pincer Search, etc.

Reinforcement Learning works by way of developing a system based on feedback from the environment, based on which required steps are taken to penalize or reward an agent for training the model. Here, learning is performed in several repetitive steps of trial and error which are based on the interaction of the agent with the surrounding environment (Shailaja et al., 2018). This process does not involve any human intervention. If the agent performs the actions as per expectation, it is rewarded, else, it is penalized (Dhillon & Singh, 2019).

Amongst the ML models, supervised learning models have been widely used for diagnosing disease using healthcare data. Designing such models involves pre-processing the given dataset followed by splitting it into training and testing data. The training data is used to train a model using supervised learning algorithms. The resultant ML-based classifier is tested using the testing data (Nasteski,2017). Based

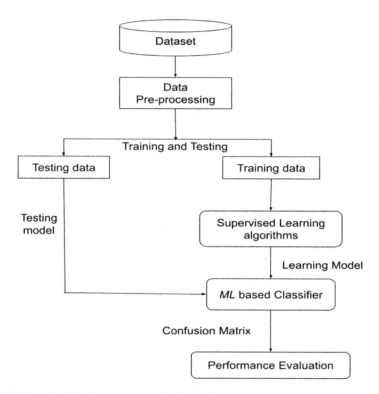

FIGURE 9.2   Classification-based machine learning.

on testing, a confusion matrix is computed using which performance evaluation of the classifier is carried out. The parameters used by the classifier are continuously tuned till an acceptable performance is achieved by the classifier. This process is depicted in Figure 9.2.

The goal of ML is to analyze healthcare data in order to assist clinicians to diagnose diseases in their early stages so that timely decisions can be taken to prevent the disease from turning acute or fatal. Further, ML would entail a lower cost in diagnosing the disease. In this paper, existing literature related to the use of ML in the diagnosis of major diseases like cancer, heart disease, diabetes and obesity have been reviewed in Section 2. Further, challenges and opportunities of the use of ML in healthcare are outlined in Section 3.

9.2   ML IN HEALTHCARE

AI is a subdivision of computer science, which has assisted in improving the quality of human life in various ways. Intelligent machines generally use AI to achieve some specific goal (Boden, 1998). AI is making healthcare diagnosis and treatment more efficient by providing better strategies and information based on data. AI techniques are being used by clinicians to make better decisions in critical areas pertaining to life threatening diseases (Yu et al., 2018). Currently, medical-image-based diagnosis using CT scans, X-rays, ultrasound, etc., are being carried out with the help of AI. Many specialists like radiologists (Reed, 2010), ophthalmologists (Panwar et al., 2016), dermatologists (Rigel et al., 2005) and pathologists (Litjens et al., 2016) make use of such image-based diagnosis. In the field of AI, ML is playing a prominent role in designing systems that provide enhanced and better-quality healthcare. ML can be used for disease detection and diagnosis, clinical monitoring of patients, framing treatment designs, drug discovery, digital consultancy, robotic surgery, etc. (Murali & Sivakumaran, 2018). It also optimizes the use of healthcare resources (Davenport & Kalakota, 2019).

ML assists clinicians in various ways and its contribution in the prediction and diagnosis of major causes of mortality has gained prime importance. According to WHO reports, heart disease, lung cancers, stroke, kidney diseases, diabetes, etc. are included in the top ten global causes of death across the world. Several research works exist in literature related to the use of ML techniques in healthcare. These include works related to cancer, cardiovascular disease, diabetes and obesity, which are the leading global causes of mortality and are briefly discussed next.

9.2.1 CANCER

Cancer is a fatal disease that results in approximately 0.3 million deaths per year. It is the second largest disease with respect to the number of deaths (Ali et al., 2011; Naresh & Shettar, 2014). Since it is a pervasive disease and is rarely detected at an early stage, there is a need of intelligent machines that can analyze data and reveal important information, which may not be possible for humans to infer manually. ML-based systems can be used in developing models that can help in the diagnosis of cancer in its early stages (Patel et al., 2019). The models could be trained using image-based data in order to predict the presence of the disease. These models can further be improved by incorporating other attributes which are directly linked with cancer, such as obesity, alcohol and tobacco consumption, genetic disorder, exposure to any radiation, monitoring food consumption, physical activities, etc. (Alani & Abbod, 2015). There are many ML techniques that can be utilized in cancer prediction, such as R, DT, SVM, ANN, LR, NB and RF.

Among the various types of cancer, lung cancer, in general, and breast cancer, in females, are more prevalent. Lung cancer is usually not detectable at early stages due to late appearance of symptoms. At an advanced stage, it becomes very difficult to cure (Patel et al., 2019). The main cause of lung cancer is the long-term smoking habit, which constitutes about 85 percent of total lung cancer cases. Other causes of lung cancer are exposure to poor air quality, genetic factors, hazardous working conditions, etc. (Taher et al., 2012). Many medical processes like CT scans, MRI scans and X-rays are used to detect it. However, these methods can only detect lung cancer at an advanced stage (Ada & Kaur, 2013). Thus, there is a need for ML techniques that can detect lung cancer at a comparatively early stage to prevent it from becoming fatal. In (Naresh & Shettar, 2014), ML techniques were utilized for early detection of stage I and stage II lung cancer. The total number of samples considered for the study were 184. Amongst these, samples pertaining to stage I and stage II were 111 and 73 respectively. SVM, ANN and k-NN (k-Nearest Neighbors) algorithms were used for lung cancer detection and evaluating its seriousness during both stage I and stage II. SVM performed the best in terms of accuracy. An Entropy Degradation Method (EDM), based on the neural network algorithm for the detection of small cell lung cancer (SCLC) using CT scans, was proposed in (Wu & Zhao, 2017). A total of 12 CT scans were taken and, amongst these, six pertained to healthy lungs and the other six were from patients of SCLC. EDM performed well in terms of accuracy.

Breast cancer is the main reason of cancer deaths among females in the fertile age range of 35–64 years. In India, one out of every two females who were diagnosed with breast cancer for the first time, dies due to it. If breast cancer is diagnosed at an early stage, mortality rates due to it can be brought down significantly. To achieve this, there is a need of computer aided technologies for early diagnosis and treatment of breast cancer (Houssami et al., 2017). In Asri et al. (2016), k-NN, DT, SVM and NB classifiers were used for diagnosis of breast cancer. SVM was found to have the highest accuracy and the lowest warning rate. In Sharma et al. (2018), various ML methods, namely RF, k-NN and NB were used for detecting breast cancer in females at an early stage. For the study, the Wisconsin Breast Cancer (WBC) dataset from the UCI repository, having 569 instances and 32 attributes, was used. k-NN performed the best in terms of detecting breast cancer. In Mohammed et al. (2020), three different classifiers, DT, NB and Sequential Minimal Optimization (SMO), were implemented on two different datasets, namely the Wisconsin Breast Cancer (WBC) dataset and the breast cancer dataset. Metrics such as accuracy, precision, recall, standard deviation, etc. were used to evaluate the performance of the models. SMO performed better in the WBC dataset while DT performed better in the breast cancer dataset. In Sharma & Mishra (2021), various ML algorithms such as k-NN, LR, DT, ANN, SVM, RF, Ada Boost (AB), etc., were used for early detection and classification of breast cancer. Various feature selection techniques such as Correlation-based Feature Selection, Sequential Feature Selection and Information Gain were considered in this study. Based on these selection approaches, features were selected, and ML techniques were applied on these. SVM performed the best followed by LR and ANN. Further, an ensemble classifier using SVM, LR and ANN performed better than the SVM classifier, LR classifier and ANN classifier.

9.2.2 CARDIOVASCULAR DISEASES

According to WHO reports, cardiovascular diseases are the topmost cause of global deaths. These diseases caused nearly 17.9 million deaths in 2016, which amount to approximately 31 percent of all global deaths. Unhealthy diet, physical inactivity, tobacco use, excessive alcohol consumption and obesity are the main behavioral risk factors associated with cardiovascular diseases (Cardiovascular Diseases, n.d.). To prevent loss of lives, it is important that these diseases are detected at an early stage. Several works have been reported in literature that use ML for cardiovascular disease prediction. In Anbarasi et al. (2010), DT and NB classifiers were used to diagnose patients with cardiovascular disease. Moreover, genetic algorithms were used for feature selection. DT was found to have the highest accuracy. In Buettner & Schunter (2019), the RF method was used to detect heart disease in patients. The model was tested with and without cross-validation. Without cross-validation, RF was found to have the highest accuracy. With 10-fold cross-validation, this accuracy increased. In Gonsalves et al. (2019), prediction of coronary heart disease using NB, SVM, DT was performed. The dataset was obtained from South African Heart Disease, which had 462 instances and ten attributes. Evaluation of the performance of ML models was based on 10-fold cross-validation. NB outperformed SVM and DT. In Bharti et al. (2021), different ML techniques such as LR, k-NN, DT, RF, SVM, XG-Boost, etc., were used for prediction of heart disease. The dataset was taken from UCI repository. On this dataset, three different approaches were used for performance improvement and evaluation. In the first approach, the entire dataset was used. SVM performed the best in terms of accuracy. In the second approach, feature selection was performed without any outlier detection. In this, RF performed the best in terms of the precision and the accuracy. In the third approach, feature selection and outlier detection were performed whereupon it was found that k-NN performed the best in terms of accuracy followed by SVM.

9.2.3 DIABETES

According to WHO, diabetes is a chronic disease in which the pancreas does not produce adequate insulin to meet the requirements of the body. The number of diabetic patients rose from 108 million in 1980 to 422 million in 2014 (Diabetes, n.d.). According to the International Diabetes Federation, 463 million people, above the age of 20 and below 80 years, suffer from diabetes (IDF Diabetes Atlas, 10th edition, n.d.). It is predicted that by the year 2045, 693 million individuals will be suffering from diabetes (Cho et al., 2018). Various ML techniques such as SVM, RF, k-NN, ANN, DT, NB, etc., have been used for diabetes detection and diagnosis (Dankwa-Mullan et al., 2019). In Aishwarya et al. (2013), a SVM classifier was used for the classification of diabetic and non-diabetic individuals. In Mahmud (2018), R, NB, RF, DT, ANN and SVM classifiers were used to predict and diagnose diabetes. In Sarwar et al. (2018), ML techniques such as NB, k-NN, SVM, LR, DT and RF were used for the prediction of diabetes. The dataset, which was taken from the UCI repository, had 768 patient records and nine attributes. Amongst all the ML techniques, k-NN and SVM provided the highest accuracy. In Sisodia & Sisodia (2018), DT, SVM and NB methods were used for early-stage detection of diabetes. The Pima Indians diabetes database from the UCI repository, a dataset of females aged 20 and above, having 768 instances with eight attributes, was considered for the study. In this, the accuracy of NB was found to be the highest. In Kaur & Kumari (2020), different ML techniques, namely Linear kernel SVM, Radial Basis Function (RBF) kernel SVM, k-NN, ANN and Multifactor Dimensionality Reduction (MDR), were used for early detection of the risk factors in diabetes. The Pima Indians diabetes dataset was considered for the study. The performance was evaluated using metrics like accuracy, precision, F1-score and Area Under ROC curve (AUC-ROC). Linear kernel SVM performed the best in terms of the accuracy and F1-score, while k-NN performed the best in terms of recall and AUC-ROC. On the Pima Indians diabetes dataset, in Reddy et al. (2020), six different ML techniques, i.e., SVM, LR, k-NN, NB, RF and Gradient Boosting (GB), were used for prediction of diabetes among individuals. Here, RF performed the best in terms of accuracy, precision, recall, F1-score and AUC-ROC. In Ghosh et al. (2021), four different ML techniques, i.e., SVM, RF, AB and GB, were used for diabetes classification using the Pima Indians diabetes dataset. Minimal Redundancy Maximal Relevance (MRMR) approach was adopted for feature selection whereupon a stratified 10-fold cross-validation approach was used. RF performed the best in terms of accuracy in both the cases, i.e., with and without feature selection.

9.2.4 OBESITY

Obesity is a chronic disease caused due to an imbalance between calorie intake and its expenditure, leading to weight gain. WHO defines obesity and overweight as the accumulation of excess fat in certain body parts leading to continuous degradation in the health of the person (De-La-Hoz-Correa et al., 2019). The conventional method of its measurement is Body Mass Index (BMI), which is calculated in terms of the weight of a person to the square of his/her height. Several works have been reported in literature related to the use of ML for predicting obesity. In Abdullah et al. (2016), various classification techniques were used for the study of obesity among children aged 12, from two districts of Terengganu; Kuala Terengganu and Besut. A total of 4,245 instances were used out of which 3,385 instances were from Kuala Terengganu and 1,692 instances were from Besut. Bayesian Network, DT (J48), ANN and SVM classifiers were used for the study and amongst these, DT performed the best in terms of predicting the risk of childhood obesity. In De-La-Hoz-Correa et al. (2019), DT (J48), NB and simple LR were used to estimate obesity levels. The data used for this study pertained to undergraduate students between the ages of 18 and 25 years. The data comprised 712 instances, out of which 324 instances pertained to men and 388 instances pertained to women. The factors included in the data to analyze behavior of individuals were age, sex, weight, diet intake, frequency of physical activity, etc. Here, DT performed the best amongst all the classifiers. In Cui et al. (2021), obesity estimation based on eating habits, physical conditions and other attributes were performed using different ML techniques. For this, a dataset titled "Estimation of obesity levels in individuals from countries of Mexico, Peru and Colombia" was considered. It has 2,111 records having 17 attributes and had individuals belonging to the age group 14 to 61 years. ML techniques such as LR, SVM, k-NN, DT, RF, XG-Boost, GB-DT and Light-GBM were applied on the whole dataset as well as the reduced dataset comprising 14 attributes. It was observed that the ensemble classifiers had better accuracy. On the same dataset, in Celik et al. (2021), estimation of obesity levels was performed using various classification techniques. DTs, such as Fine Tree, Medium Tree, Coarse Tree, Linear Tree, Quadratic Tree, Boosted Trees, Bagged Trees and RUS Boosted Trees; and SVM, such as Cubic SVM, Fine SVM, Medium Gaussian SVM, Coarse SVM were applied on the dataset. Amongst these ML techniques, the Cubic SVM performed the best.

TABLE 9.1
Use of Various ML Techniques for Disease Diagnosis

Diseases	Data Source	Method (s)	Reference
Thyroid diseases	UCI	ANN	Gharehchopogh et al., 2013
	UCI	k-NN, DT, SVM, ANN	Tyagi et al., 2018
Mental illness	Questionnaire based	k-NN, DT, RF, NB, LR, SVM	Srividya et al., 2018
Blood disorder	Hospital records	ANN	Payandeh et al., 2009
Hepatitis	UCI	NB, DT, RF, ANN	Karthikeyan & Thangaraju, 2013
Sleep disorder	Clinical records	LR	Cai et al., 2017
Pregnancy disorders	Hospital records	ANN	Maylawati et al., 2017
	-	NB	Moreira et al., 2016

Table 9.1 shows detail of some of the works where ML techniques have also been used on datasets related to diseases such as thyroid, mental illness, blood disorder, hepatitis, sleep disorder and pregnancy disorder.

The ultimate goal of ML techniques is to design healthcare systems that would enable progression from reactive to proactive healthcare. However, there are several challenges and opportunities in designing such systems. These are discussed next.

9.3 CHALLENGES AND OPPORTUNITIES FOR ML IN HEALTHCARE

A healthcare system that aims to ensure the proactive care of individuals should have the capability to capture, store, access, analyze and visualize large amounts of healthcare data. This poses several challenges and opportunities, which are briefly discussed below:

- The availability of healthcare data is of prime importance and therefore appropriate tools and technology should be used to capture this data. Further, while capturing data relevant to a disease, features having major impact in causing the disease should be identified and data related to these should be stored. ML can help in identifying correlations of a feature with the disease using dimensionality reduction techniques.
- For proactive healthcare, data of past patients with labels specifying the presence or absence of a disease and in case, if it is present, the level of severity of disease is required. However, past data is available but is not labelled and therefore it becomes difficult in diagnosing a disease. ML techniques can be used to deduce key and insightful patterns in the data that would help in diagnosing a disease.
- Clinical data is very essential for effective diagnosis of a disease. However, it is not available in digitized form and is recorded manually by a physician based on his/her interaction with the patients. This may not be completely accurate due to human error. This data needs to be collected and digitized in real time

and ML can help in real time diagnoses based on the healthcare records and clinical tests of patients.

- For any drug or vaccine development, clinical trial data needs to be continuously monitored and digitized. Moreover, analyzing this data consumes a lot of time, and decisions based thereupon may not always be accurate. ML can help in identifying the right individuals for trials and can help in observing their responses more effectively and efficiently. Also, ML can help in developing personalized medicine by analyzing the sensitivity of the patient towards the medicine.

- Usually, clinical records are heterogeneous, noisy and may have a large number of features. Some of these features may not be relevant for diagnosing a disease and some may be inter-related. Further, due to inconsistency in capturing this data, values for features may be missing. Also, the data captured could be imbalanced as only a small number of data instances may be related to individuals having the disease. These missing values and class imbalances in the data can also affect the performance of the ML classifiers. Therefore, there is a need to perform data pre-processing, which may include computing missing values from other data instances; and using class-balancing techniques to balance the dataset.

- The healthcare data generated is usually semi-structured and/or unstructured. Therefore, there is a challenge in accessing and analyzing such data. Further, as part of medical diagnosis, laboratory tests may involve capturing imagery data such as X-Ray/CT scan/MRI images of infected organs. Even the most experienced physicians can overlook minor details in the image, which may be key for early diagnosis of a disease. Deep Learning (DL), a branch of ML, can help in deriving useful insights consisting of the minutest of details from this image data. DL techniques such as recurrent neural network and convolution neural network can be used for comprehending and analyzing such data.

- Epidemics and pandemics are rare events but whenever they occur, they cause severe damage to a country, a continent and sometimes the entire world. These are health disasters that impact many lives and livelihoods. They also ruin the economy of a country. The spread of diseases that cause epidemics or pandemics can be curtailed by using machine learning techniques. ML can help in recognizing and predicting the spread of infectious diseases. It can help in identifying hotspots of the disease using unsupervised ML techniques so as to enable healthcare workers and government agencies to develop plans accordingly that would prevent the disease from spreading further.

- Proactive healthcare measures should result in transition from predictive to preventive healthcare, so as to prevent the disease from turning severe. ML can help in predicting patients whose health condition is likely to become critical so that preventive and proactive measures can be taken to provide timely and requisite medical care and facilities during the early stages. ML can also be used to classify patients in terms of their severity condition so that better preparedness in terms of the required medical infrastructure and facilities can be arranged.

- Generalization of classification results is a challenging task. For instance, in case of rare diseases, the data available may be insufficient. Analysis based on this data could lead to ineffective models, which in turn may give results from which no useful inference can be drawn. Transfer learning, a type of ML, can be used to design a generic domain-specific model that can be applied on the rare disease data for effective diagnosis of the disease.
- In the healthcare domain, privacy and security of personal medical data is of prime importance. A breach of this may lead to severe psychological and mental distress to the patient. ML techniques can be used to design strategies that preserve the security and integrity of personal data.

9.4 CONCLUSION

ML has emerged as a powerful tool to improve the ability, as well as performance, of decision making in many fields, with healthcare being one among them. It has even surpassed the ability, efficiency and performance of humans involved in the healthcare sector in various subdomains such as the diagnosis of disease, treatments, medicines, etc. There is a rapid technological revolution undergoing in this sector due to the emergence of ML. ML is able to deduce insightful information from healthcare data and is thereby helping clinicians to take better informed and timely decisions in order to provide better care, at effective cost to save lives. The existing works in this area have shown that the underlying technological revolution has helped in the detection, diagnosis and treatment of diseases causing global deaths such as cancer, diabetes, cardiovascular diseases, obesity, etc. ML is providing resourceful avenues in a new era of preventing, monitoring and treating life-threatening diseases. In addition, it has the capability to support inexperienced clinicians in different healthcare domains where specialist physicians may not be easily available. Even routine tasks can be effectively handled using ML, as there are more chances of human error in performing repetitive tasks. The use of ML would not only strengthen the healthcare services but would also boost preventive care and early detection of disease to save many lives. DL is now being widely used in various subdomains of healthcare that involve medical-image data related to ophthalmology, cardiology, dermatology, pathology, etc. Further, there is a need of multidisciplinary and multi-sector collaborations to accelerate the development as well as deployment of ML-based healthcare systems. Though these technologies are efficient in many domains, they cannot completely substitute human physicians. Instead, these can assist clinicians in making better informed and timely decisions with greater accuracy and speed.

REFERENCES

Abdullah, F. S., Manan, N. S. A., Ahmad, A., Wafa, S. W., Shahril, M. R., Zulaily, N., Amin, R. M., & Ahmed, A. (2016). Data mining techniques for classification of childhood obesity among year 6 school children. *Advances in Intelligent Systems and Computing*, 465–474. doi:10.1007/978-3-319-51281-5_47

Ada & Kaur, R. (2013). Using some data mining techniques to predict the survival year of lung cancer patient. *International Journal of Computer Science and Mobile Computing*, 2(4), 1–6.

Alani, S., & Abbod, M. (2015). Performance of cancer prediction based on artificial neural network. doi: 10.17758/ur.u1214024.

Ali, I., Wani, W. A., & Saleem, K. (2011). Cancer scenario in India with future perspectives. *Cancer Therapy*, 8, 56–70.

Aishwarya, R., Gayathri, P., & Jaisankar, N. (2013). A method for classification using machine learning technique for diabetes. *International Journal of Engineering and Technology*, 5(3), 2903–2908.

Anbarasi, M., Anupriya, E., & Iyengar, N. C. S. N. (2010). Enhanced prediction of heart disease with feature subset selection using genetic algorithm. *International Journal of Engineering Science and Technology*, 2(10), 5370–5376.

Asri, H., Mousannif, H., Moatassime, H. A., & Noel, T. (2016). Using machine learning algorithms for breast cancer risk prediction and diagnosis. *Procedia Computer Science*, 83, 1064–1069. doi: 10.1016/j.procs.2016.04.224.

Bhardwaj, R., Nambiar, A. R., & Dutta, D. (2017). A study of machine learning in healthcare. 2017 IEEE 41st Annual Computer Software and Applications Conference (COMPSAC), 236–241. doi:10.1109/COMPSAC.2017.164.

Bharti, R., Khamparia, A., Shabaz, M., Dhiman, G., Pande, S., & Singh, P. (2021). Prediction of heart disease using a combination of machine learning and deep learning. *Computational Intelligence and Neuroscience*, 2021, 1–11. doi:10.1155/2021/8387680.

Boden, M. A. (1998). Creativity and artificial intelligence. *Artificial Intelligence*, 103(1–2), 347–356. doi:10.1016/s0004–3702(98)00055-1.

Buettner, R., & Schunter, M. (2019). Efficient machine learning based detection of heart disease. 2019 IEEE International Conference on E-health Networking, Application & Services (HealthCom). doi:10.1109/healthcom46333.2019.9009429.

Cai, L., Chen, C., Wang, X., Yang, X., Lin, S., Huang, J., Jiang, J., Datta, R., Du, M., Jiang H., Zhu, M., & Huang, J. (2019). Sleep disorder classification method based on logistic regression with apnea-ECG dataset. In Proceedings of the 2019 International Conference on Artificial Intelligence and Advanced Manufacturing, 1–4. https://doi.org/10.1145/3358331.3358344

Cardiovascular Diseases. (n.d.). World Health Organization. Retrieved October 10, 2021, from https://www.who.int/health-topics/cardiovascular-diseases#tab=tab_1

Celik, Y., Guney, S., & Dengiz, B. (2021). Obesity level estimation based on machine learning methods and artificial neural networks. 2021–44th International Conference on Telecommunications and Signal Processing (TSP). doi:10.1109/tsp52935.2021.9522628.

Cho, N., Shaw, J., Karuranga, S., Huang, Y., Fernandes, J. D., Ohlrogge, A., & Malanda, B. (2018). IDF diabetes atlas: Global estimates of diabetes prevalence for 2017 and projections for 2045. *Diabetes Research and Clinical Practice*, 138, 271–281. doi: 10.1016/j.diabres.2018.02.023.

Cui, T., Chen, Y., Wang, J., Deng, H., & Huang, Y. (2021). Estimation of obesity levels based on decision trees. 2021 International Symposium on Artificial Intelligence and Its Application on Media (ISAIAM). doi:10.1109/isaiam53259.2021.00041.

Dankwa-Mullan, I., Rivo, M., Sepulveda, M., Park, Y., Snowdon, J., & Rhee, K. (2019). Transforming diabetes care through artificial intelligence: The future is here. *Population Health Management*, 22(3), 229–242. doi:10.1089/pop.2018.0129.

Davenport, T., & Kalakota, R. (2019). The potential for artificial intelligence in healthcare. *Future Healthcare Journal*, 6(2), 94–98. https://doi.org/10.7861/futurehosp.6-2-94.

De-La-Hoz-Correa, E., Mendoza-Palechor, F. E., De-La-Hoz-Manotas, A., Morales-Ortega, R. C., & Adriana, S. H. (2019). Obesity level estimation software based on decision trees. *Journal of Computer Science*, 15(1), 67–77. doi:10.3844/jcssp.2019.67.77.

Dhillon, A., & Singh, A. (2019). Machine learning in healthcare data analysis: A survey. *Journal of Biology and Today`s World*, 8, 1–10. doi: 10.15412/J.JBTW.01070206.

Diabetes. (n.d.). Retrieved October 10, 2021, from https://www.who.int/health-topics/diabetes#tab=tab_1

Firdaus, H., & hassan, S. I. (2020). Unsupervised learning on healthcare survey data with particle swarm optimization. *Learning and Analytics in Intelligent Systems Machine Learning with Health Care Perspective*, 57–89. doi:10.1007/978-3-030-40850-3_4.

Gharehchopogh, F. S., Molany, M., & Mokri, F. D. (2013). Using artificial neural network in diagnosis of thyroid disease: a case study. *International Journal on Computational Sciences & Applications* 3(4), 49–61.

Ghosh, P., Azam, S., Karim, A., Hassan, M., Roy, K., & Jonkman, M. (2021). A Comparative study of different machine learning tools in detecting diabetes. *Procedia Computer Science*, 192, 467–477. doi:10.1016/j.procs.2021.08.048.

Ginneken, B. V., Setio, A. A. A., Jacobs, C., & Ciompi, F. (2015). Off-the-shelf convolutional neural network features for pulmonary nodule detection in computed tomography scans. 2015 IEEE 12th International Symposium on Biomedical Imaging (ISBI). doi: 10.1109/isbi.2015.7163869.

Gonsalves, A. H., Thabtah, F., Mohammad, R. M., & Singh, G. (2019). Prediction of Coronary Heart Disease using Machine Learning. Proceedings of the 2019–3rd International Conference on Deep Learning Technologies – ICDLT 2019. doi:10.1145/3342999.3343015.

Houssami, N., Lee, C. I., Buist, D. S., & Tao, D. (2017). Artificial intelligence for breast cancer screening: Opportunity or hype? *The Breast*, 36, 31–33. doi: 10.1016/j.breast.2017.09.003.

IDF Diabetes Atlas 10th Edition. (n.d.). Retrieved October 22, 2021, from https://www.diabetesatlas.org/en/

Karthikeyan, T., & Thangaraju, P. (2013). Analysis of classification algorithms applied to hepatitis patients. *International Journal of Computer Applications,* (0975 – 888), 2(15).

Kaur, H., & Kumari, V. (2020). Predictive modelling and analytics for diabetes using a machine learning approach. *Applied Computing and Informatics*, doi:10.1016/j.aci.2018.12.004.

Litjens, G., Sánchez, C. I., Timofeeva, N., Hermsen, M., Nagtegaal, I., Kovacs, I., Hulsbergen, C., Bult, P., Ginneken, B.V., & Laak, J.V.D., (2016). Deep learning as a tool for increased accuracy and efficiency of histopathological diagnosis. *Sci Rep* 6, 26286. https://doi.org/10.1038/srep26286

Mahesh, B. (2020). Machine learning algorithms: A review. *International Journal of Science and Research*, 9, 381–386.

Mahmud, S. M., Hossin, M. A., Ahmed, M. R., Noori, S. R., & Sarkar, M. N. (2018). Machine learning based unified framework for diabetes prediction. Proceedings of the 2018 International Conference on Big Data Engineering and Technology – BDET 2018. doi:10.1145/3297730.3297737.

Maylawati, D. S. A., Ramdhani, M. A., Zulfikar, W. B., Taufik, I., & Darmalaksana, W. (2017). Expert system for predicting the early pregnancy with disorders using artificial neural network. In 2017–5th International Conference on Cyber and IT Service Management (CITSM), 1–6, doi: 10.1109/CITSM.2017.8089243.

Mohammed, S. A., Darrab, S., Noaman, S. A., & Saake, G. (2020). Analysis of breast cancer detection using different machine learning techniques. *Data Mining and Big Data Communications in Computer and Information Science*, 108–117. doi:10.1007/978-981-15-7205-0_10.

Moreira, M. W., Rodrigues, J. J., Oliveira, A. M., Saleem, K., & Neto, A. V. (2016). An inference mechanism using bayes-based classifiers in pregnancy care. In 2016 IEEE

18th International Conference on e-Health Networking, Applications and Services (Healthcom), 1–5, doi:10.1109/HealthCom.2016.7749475.

Murali, N., Sivakumaran, N. (2018). Artificial intelligence in healthcare: A review. *International Journal of Modern Computation, Information and Communication Technology*, 1(6): 103–110.

Naresh, P., & Shettar, D. R. (2014). Early detection of lung cancer using neural network techniques. *Prashant Naresh Int. Journal of Engineering Research and Applications*, ISSN: 2248–9622, 4(8), 78–83.

Nasteski, V. (2017). An overview of the supervised machine learning methods. *Horizons.b*, 4, 51–62. doi:10.20544/horizons.b.04.1.17.p05.

Osisanwo, F. Y., Akinsola, J. E. T., Awodele, O., Hinmikaiye, J. O., Olakanmi, O., & Akinjobi, J. (2017). Supervised machine learning algorithms: Classification and comparison. *International Journal of Computer Trends and Technology*, 48(3), 128–138. doi:10.14445/22312803/ijctt-v48p126.

Patel, D., Shah, Y., Thakkar, N., Shah, K., & Shah, M. (2019). Implementation of artificial intelligence techniques for cancer detection. *Augmented Human Research*, 5(1). doi:10.1007/s41133–019–0024–3.

Panwar, N., Huang, P., Lee, J., Keane, P. A., Chuan, T. S., Richhariya, A., Teoh, S., Lim, T. H., & Agrawal, R. (2016). Fundus photography in the 21st century – a review of recent technological advances and their implications for worldwide healthcare. *Telemedicine and e-Health*, 22(3), 198–208. https://doi.org/10.1089/tmj.2015.0068

Payandeh, M., Aeinfar, M., Aeinfar, V., & Hayati, M. (2009). A new method for diagnosis and predicting blood disorder and cancer using artificial intelligence (artificial neural networks). *International Journal of Hematology-Oncology and Stem Cell Research*, 3(4), 25–33.

Reed, J. C. (2010). *Chest Radiology Plain Film Patterns and Differential Diagnoses E-Book*. United Kingdom: Elsevier Health Sciences.

Reddy, J. D., Mounika, B., Sindhu, S., Pranayteja Reddy, T., Sagar Reddy, N., Jyothsna Sri, G., Swaraja, K., Meenakshi, K., & Kora, P. (2020). Predictive machine learning model for early detection and analysis of diabetes. *Materials Today: Proceedings*. https://doi.org/10.1016/j.matpr.2020.09.522

Rigel, D. S., Friedman, R. J., Kopf, A. W. & Polsky, D., (2005). ABCDE – an evolving concept in the early detection of melanoma. *Arch. Dermatol.* 141, 1032–1034.

Sarwar, M. A., Kamal, N., Hamid, W., & Shah, M. A. (2018). Prediction of diabetes using machine learning algorithms in healthcare. 2018–24th International Conference on Automation and Computing (ICAC). doi:10.23919/iconac.2018.8748992.

Shailaja, K., Seetharamulu, B., & Jabbar, M. A. (2018). Machine learning in healthcare: A Review. 2018 Second International Conference on Electronics, Communication and Aerospace Technology (ICECA). https://doi.org/10.1109/iceca.2018.8474918

Sharma, A., & Mishra, P. K. (2021). Performance analysis of machine learning based optimized feature selection approaches for breast cancer diagnosis. *International Journal of Information Technology*. doi:10.1007/s41870–021–00671–5.

Sharma, S., Aggarwal, A., & Choudhury, T. (2018). Breast Cancer Detection Using Machine Learning Algorithms. 2018 International Conference on Computational Techniques, Electronics and Mechanical Systems (CTEMS). doi:10.1109/ctems.2018.8769187.

Sisodia, D., & Sisodia, D. S. (2018). Prediction of diabetes using classification algorithms. *Procedia Computer Science*, 132, 1578–1585. doi:10.1016/j.procs.2018.05.122.

Srividya, M., Mohanavalli, S., & Bhalaji, N. (2018). Behavioral modeling for mental health using machine learning algorithms. *Journal of medical systems*, 42(88). https://doi.org/10.1007/s10916-018-0934-5.

Taher, F., Werghi, N., & Al-Ahmad, H. (2012). Bayesian classification and artificial neural network methods for lung cancer early diagnosis. 2012–19th IEEE International Conference on Electronics, Circuits, and Systems (ICECS 2012). doi:10.1109/icecs.2012.6463545.

Tekkeşin, A. I. (2019). Artificial intelligence in healthcare: Past, present and future. *The Anatolian Journal of Cardiology*. doi:10.14744/anatoljcardiol.2019.28661.

Tyagi, A., Mehra, R., & Saxena, A. (2018). Interactive thyroid disease prediction system using machine learning technique. In 2018 Fifth International Conference on Parallel, Distributed and Grid Computing (PDGC), 689–693, IEEE.

Wu, Q., & Zhao, W. (2017). Small-cell lung cancer detection using a supervised machine learning algorithm. 2017 International Symposium on Computer Science and Intelligent Controls (ISCSIC). doi:10.1109/iscsic.2017.22.

Yu, K., Beam, A. L., & Kohane, I. S. (2018). Artificial intelligence in healthcare. *Nature Biomedical Engineering*, 2(10), 719–731. doi:10.1038/s41551–018–0305-z.

10 Transfer Learning and Fine-Tuning-Based Early Detection of Cotton Plant Disease

Himani Bhatheja
Department of Electronics and Communication
Engineering, Delhi Technological University,
Delhi, India

N. Jayanthi
Assistant Professor, Department of Electronics and
Communication Engineering, Delhi Technological
University, Delhi, India

CONTENTS

DOI: 10.1201/9781003279044-10

10.1 INTRODUCTION

Cotton plant production is very crucial for our country. Cotton is used in various industries and therefore its production should be maintained and supervised with modern techniques which can help to increase the throughput and ultimately increase in the economy of our country. Cotton crops can suffer from some diseases like Cercospora, bacterial blight, Ascochyta blight, and target spot. Thus, identifying these diseases at an early time becomes crucial for the farmer to produce a healthy harvest.

Modern technologies and advancement can give the famers a means to prevent damage happening to their crops. Generally, conventional machine-learning algorithms such as K-means clustering and support vector machine have complex image processing and feature extraction methods, which reduces the accuracy of disease identification. These techniques are better suited for the detection of plant images in consistent background which have been performed in a consummate lab condition (Chung et al., 2016). There have been various advancements in research in the Deep Learning field to predict diseases at an early stage in the cotton plants and leaves. For example, transfer learning which is very useful in the field of Computer Vision, Image Processing and Natural Language Processing particularly due to following reasons:

- It needs less training data
- It helps models generalizes better
- It makes training easier to debug

Transfer Learning is a method where a deep convolutional neural network is first trained on a problem which is somewhat similar to a problem that is to be solved. Then, one or more layers from the pre-trained model are used in the network which is being trained on a problem statement that is of interest. It is basically the sum of the convolutional neural networks designed, which are actually producing state-of-the-art algorithms to classify different kinds of images.

Figure 10.1 describes the basic understanding of transfer learning that is using the previously trained model, i.e., the old classifier, for a new type of problem statement, i.e., the new classifier. The proposed work consists of analysis of two of these techniques, Inception V3 and EfficientNet B0, used along with fine-tuning techniques to enhance the detection of disease in cotton plant leaves. The insights drawn from the implementation on two datasets led to a better understanding of the effect of fine-tuning on pre-trained convolutional neural networks. Data preprocessing such as data augmentation techniques were used to better train the models. Publically available datasets were used for this purpose.

In this paper, a summary of previous related work will be given in the form of a literature review in Section 10.2. The transfer learning methods, the image preprocessing work and the fine-tuning techniques used will be introduced in Section 10.3. The steps performed during implementation are depicted in Section 10.4. The experimental results using various metrics are given followed by conclusive comments and a summary in Section 10.5 and 10.6.

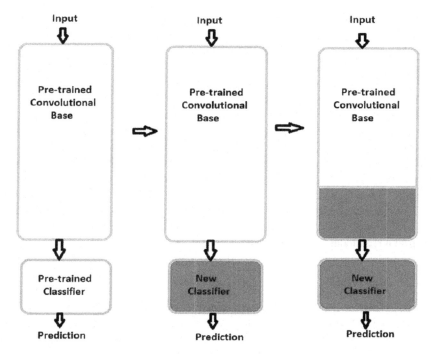

FIGURE 10.1 Understanding transfer learning.

10.2 RELATED WORK

Chia-Lin Chung et al. (2016) used Support Vector Machine (SVM) to build a classifier to detect healthy or diseased seedling in rice. After optimizing parameters, the model was able to classify infected seedling with 87.6% accuracy. The authors claimed that this automated approach is more superior than manual detection.

Sourabh Shrivastava et al. (2017) presented another technique to reduce manual detection of diseases by employing SVM, K-Nearest Neighbors (KNN) and Probabilistic Neural Network (PNN) to detect six types of diseases in soya bean plants. They tested several colors and texture-based feature descriptors and emphasized its importance in machine learning models.

Liu et al. (2016) used SVM classifier to detect aphids in wheat-related images. They mentioned that features such as color, texture, density, weather and location had immense impact on detection rate of aphids in wheat. Akhtar et al. (2013) segmented the images from the background and then different types of handcrafted features are extracted from the images which are later used in different machine learning models such as KNN, Naïve Bayes, Decision Trees and SVM.

Chen et al. (2020) used VGGNet pre-trained on the ImageNet and Inception modules by using pre-trained weights. A network generated by replacement layers consisted of pre-trained layers that was used in feature extraction and an additional structure which is used for the purpose of classification. They achieved a validation accuracy of 91.83% on publically available datasets.

The paper from Too et al. (2019) describes a comparative study on different transfer learning architectures without data augmentation. The dataset used was PlantVillage with 14 types of plant leaf images. The authors mentioned that DenseNet has ability to improve the accuracy with increasing number of epochs without overfitting and deteriorating the performance. Finally, the test accuracy of DenseNet was 99.75% with a drawback of large computational time.

In the paper by Prajwala et al. (2018), the authors have used a model that has convolution and max pooling layers following two fully connected layers on Tomato plant disease dataset. They implemented various filtering techniques in each layer and used data augmentation for balancing the images inside each class. The test accuracy ranged between 76% to 100% and average testing accuracy is 91.2%.

Guan Wang et. al. (2017) used deep CNN on the PlantVillage dataset to identify the disease complexity in apple rot images of four severity stages. They evaluated the performance of deep and shallow network models. Finally, it was mentioned that deep VGG16 model is best, providing accuracy of 90.4%.

10.3 METHODS

The Convolutional Neural Networks consists of multiple layers: convolutional, pooling and fully connected layers, in which the first two layers are used for extracting features from input images and the last layer is used for classification (Akhtar et al., 2013). Pooling layer is used to decrease the dimensionality of extracted attributes. Softmax layer is used with the fully connected layer to perform the classification.

The entire method of detection using a model is a step-by-step process devised into a number of stages that are:

10.3.1 Data Preprocessing

The datasets of diseased and fresh cotton leaves were downloaded from the cotton disease dataset. After collection of datasets, the images went through preprocessing and after that the images were trained using the model. For different deep learning techniques, data preprocessing includes image filtering and dataset preparation (Chung et al., 2016). All the images were resized, and data augmentation was done on a training set to get a wider variety of images for training purpose. A horizontal shift along with height shift, width shift, zoom range and rotation value of 0.2, 0.2, 0.2 and 0.2 were given in order to perform augmentation.

10.3.2 Fine-Tuning

The transfer learning models were trained on a large dataset (ImageNet) and there-fore used as generalized model for training wide variety of images. These models were improved for a dedicated problem using fine-tuning. Generally, it is good prac-tice to take raw samples of a small dataset, with augmentation done on them and get better results from fine-tuning. Fine-tuning can be done with one or combination of following techniques (Figure 10.2):

Original fresh cotton plant image : Shape : (694, 694, 3)   Augmented fresh cotton plant image

Shape: (299, 299, 3)

FIGURE 10.2   Original image and augmented image.

A common technique is to shorten final layer which is softmax (generally) and replace it with a new combination of convolutional layers which is called new softmax layer which is in relevance to the specific task. For instance, the pre-trained model implemented on the ImageNet has 1000 categories in softmax layer and our task is dedicated to classifying 20 categories, then a newly formed softmax layer could be modified for 20 categories. Back propagation can be then used to fine tune the pre-trained weights.

A smaller learning rate is used in order to fit the model. Learning rate is a hyperparameter used to control how quickly a network learns. A big value of learning rate can make a network learn fast but with not A smaller value of learning rate can make a model learn optimally but can take a longer time. So, it should not be too large nor too small. The value of learning rate used while developing the pre-trained network could be set to a lower value in order to optimally train weights for dedicated task.

Another method to fine-tune a network is to freeze the weights of some initial layers of model we wish to use. It is done to use first a few layers for capturing general features such as edges and curves which are normally relevant to every problem. By keeping those weights, the focus of network can be shifted to learn dataset specific features in following layers.

In this experiment, a combination of all three fine-tuning techniques is used. The base models are used as Inception V3 and EfficientNet B0, as shown in figure 10.3, along with data augmentation layer. Then, new softmax layer is added and a learning rate with value 0.001 is given for training the dataset generation (Szegedy et al., 2016).

10.3.3   INCEPTIONV3

GoogleNet module was the initialization of Inception CNN architecture. It is good to assist in object detection and image analysis. In ImageNet Recognition Challenge, the Google's Inception convolutional neural network was introduced. InceptionV3 is considered to be a third version of the original architecture of version 1. InceptionV3 is very swift and also accurate in performing as CNN classifiers (Szegedy et al., 2016). There have been several developments in breaking out the different versions of Inception model, as this is one of major state-of-the-art models that can be used efficiently for image classification. Inception module architecture uses concatenation of feature maps generated by kernels of varying dimensions.

The Inception model architecture is more of a wider approaching architecture rather than deeper. It has a total of 310 layers. It promotes this type of architecture as accuracy doesn't increase just by increasing the depth of convolutional neural networks. Therefore, different versions of the Inception model introduce different components of filter concatenation and/or layers used.

10.3.4   EFFICIENTNET

Mingxing Tan and Quoc V. Le proposed the Efficient model in 2019. The researchers studied model scaling and discovered that a deep network can be scaled in three ways: depth, width and resolution. The depth can be increased by adding more layers, width can be increased by adding layers' widthwise, and resolution can be scaled

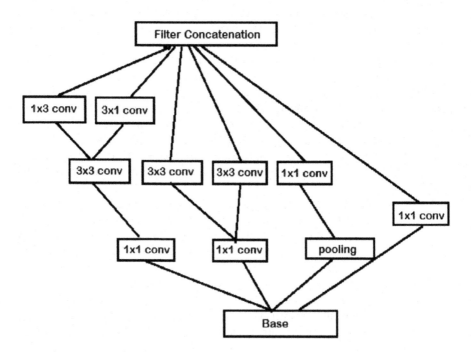

FIGURE 10.3   Basic module of inception V3.

by increasing resolution of the input to find fine-grain features of the input. The layer architecture is not changed in scaling, and only width, depth and resolution is changed. They found that careful balancing of these scaling dimensions can lead to better performance (Tan & Le, 2019).

A scaling technique was introduced that uniformly scales width, depth and resolution of the network. Compound scaling is done by performing grid search as first step to find relationship between different scaling dimensions that of the baseline network with constraints of fixed resource. Appropriate scaling coefficients are then determined for all dimensions. Then these coefficients are applied to scale up the baseline model to the desired target network size.

This method of compound scaling improves model efficiency and accuracy consistently for scaling up pre-trained models like ResNet rather than conventional scaling techniques.

The authors have proposed to use the Neural Architecture Search (NAS)-based new baseline network using AutoML MNAS framework. Mobile inverted bottleneck convolution (MBConv) architecture is used (Szegedy et al., 2016), as shown in Figure 10.4, and scaling is then used to get a group of deep learning convolutional models, called EfficientNets, which are more accurate and efficient in comparison to the earlier CNNs (Szegedy et al., 2016).

To test efficiency of the model, they also used transfer learning to test for different datasets than ImageNet and they all outperformed the previous models in 5 out of 8 datasets, suggesting that it is efficient for transfer learning also.

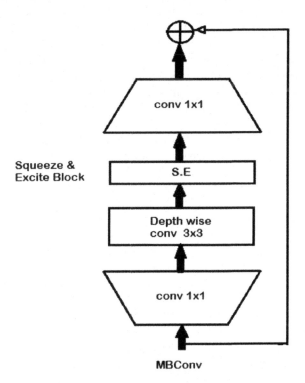

FIGURE 10.4 Mobile inverted bottleneck (MBConv) architecture.

10.4 IMPLEMENTATION

10.4.1 HARDWARE AND SOFTWARE SETUP

The experiments were performed on a Windows 10 PC with 64- bit operating system. The model training was done using 12 GB NVIDIA Tesla K80 GPU. The software tools used were Keras 2.2.6, Tensorflow 1.13.0, Matplotlib 3.4.1 and Python 3.8.2.

10.4.2 IMAGE ACQUISITION

The dataset is collected from the cotton disease dataset with four classes for diseased cotton leaves, diseased cotton plant, fresh cotton leaves and fresh cotton plants.

10.4.3 DATA PREPROCESSING AND AUGMENTATION

The images are rescaled to 224 × 224 pixels. Different augmentation techniques like rotation, height shifting, width shifting, zooming the image, and horizontal shifting are applied to the training set using ImageDataGenerator from Keras to generate new images for training.

10.4.4 TRAINING

The dataset is loaded for training, testing and validation. The total parameters for Inception V3 and EfficientNetB0 are 21,810,980 and 4,054,695, and the total number of layers are 310 and 236 respectively. For every experiment, categorical cross-entropy and accuracy metric are used for evaluation of the models. The performance of both the models before and after fine-tuning is shown in Table 10.1. The metrics used for comparison are accuracy, loss, F1 score, recall and precision.

Another dataset is also fed to the model for testing it for different cotton images taken from internet and resized. The performance of model for these images is also listed in Tables 10.1 and 10.2. Each experiment runs for 25 epochs before fine-tuning and additional 15 epochs after fine-tuning. Hyper-parameters for the experiments were standardized on both model networks. Stochastic Gradient Descent (SGD) is used in training for increasing the training speed and easy convergence. Batch size is kept at 32 to achieve better training stability. For all model networks, the learning rate value was taken as 0.001.

10.5 EXPERIMENTS AND RESULTS

During this research, the effect of fine-tuning on pre-trained convolutional networks was observed using cotton plant disease detection. The focus was using data augmentation and fine-tuning to enhance the performance of pre-trained models. Overfitting can happen because the size of training set is less. Therefore, the model tends to overfit on a small dataset because a large variety of data is not being provided for training. With data augmentation, the model training is done effectively as we are changing the width, height, and applying rotation, zoom and horizontal shift. Figure 10.5 shows overview of implementation steps.

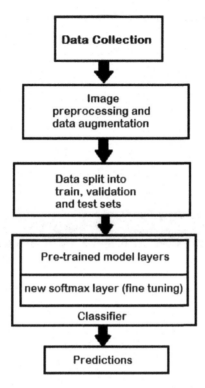

FIGURE 10.5 Overview of implementation steps.

The performance of Inception V3 is shown in Figure 10.6. The model was fit on 1951 images for 35 epochs. Figure 10.6 shows the plot of accuracy against epochs on top, and loss against epochs on bottom. The green line is used to depict the result of fine-tuning on performance of both the networks. The accuracy and other metrics are depicted in Table 10.1 and 10.2.

As shown in Figure 10.6 for InceptionV3 implementation, it was observed that Validation Loss was fluctuating very much if no fine-tuning was done on the dataset. After introducing fine-tuning, the loss decreased further, and accuracy increased and became more stable.

Performance of EfficientNet B0 is similarly shown in Figure 10.7 for comparison of effect of fine-tuning on different types of diseased cotton plant images. The validation and training accuracy graphs of the model before and after fine-tuning depicts that the accuracy improved significantly after introducing a fine-tuning method.

The proposed networks are tested on 359 images from the dataset using the methodology described in Section 10.3. Tables 10.1 and 10.2 depict the results according to different metrics. After fine-tuning, the accuracy of model on validation data has increased for both models.

Inception V3 performs poorly with overall validation accuracy of 69.13% before tuning. This accuracy is increased, after fine-tuning is performed, to 78.12%.

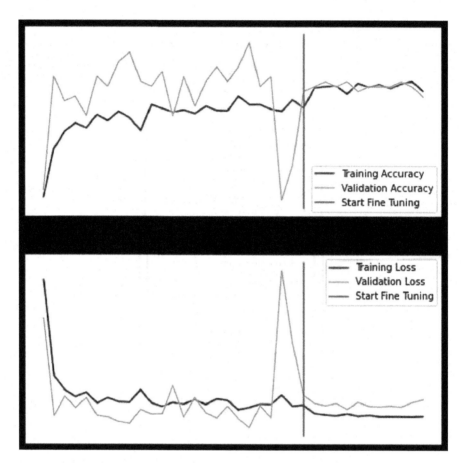

FIGURE 10.6   Inception V3 with fine-tuning.

EfficientNetB0 performs adequately with overall validation accuracy of 79.84% before tuning, and 88.93% after fine-tuning.

Overall, it is observed that EfficientNetB0 performed better than Inception V3 for diseased cotton plant images. The execution time for EfficientNet B0 is much less compared to the Inception V3 and F1 score. Precision and recall are also better for EfficientNetB0.

10.6   CONCLUSION AND FUTURE SCOPE

In this paper, state-of-the-art deep convolutional networks are evaluated for identification of diseased cotton plant identification using fine-tuning. The deep neural network methods used in this work are EfficientNet B0 and Inception V3 using transfer learning technique. By using data augmentation, better training of data is done as it increases the quantity and diversity of data used in training. EfficientNet architecture uses compound scaling and mobile inverted bottleneck convolution which has

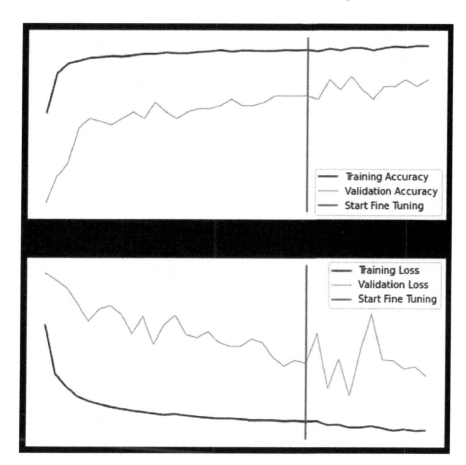

FIGURE 10.7   Performance of efficientNet B0.

TABLE 10.1
Results Before Fine-Tuning

Metrics	Results	
	InceptionV3	EfficientNetB0
Train Accuracy	74.27%	95.75%
Validation Accuracy	**69.13%**	**79.84%**
Execution Time	1001 sec	796 sec

performed better than Inception module architecture that used concatenating feature maps generated by kernels of varied dimensions. For fine-tuning, a lower learning rate and new softmax layer are used that give an accuracy of 88.93% for EfficientNetB0 and 79.84% for InceptionV3. EfficientNet model implemented with fine-tuning is

TABLE 10.2
Results After Fine-Tuning

Metrics	Results	
	InceptionV3	EfficientNetB0
Train Accuracy	82.93%	97.69%
Validation Accuracy	**78.12%**	**88.93%**
Execution Time	1356 sec	984 sec
Precision	0.76	0.91
F1 Score	0.66	0.90
Recall	0.66	0.90

more accurate for cotton plant disease detection than the Inception model and also takes less computational time.

In future, we will introduce more architectures using transfer learning on a larger dataset with more types of diseases of cotton plant that will facilitate farmers to prevent cotton crops from damage.

REFERENCES

Akhtar, A., Khanum, A., Khan, S.A., & Shaukat, A. (2013). Automated Plant Disease Analysis (APDA): Performance Comparison of Machine Learning Techniques. *2013–11th International Conference on Frontiers of Information Technology*, 60–65.

Chen Junde, Chen Jinxiu, Zhang Defu, Sun, Yuandong & Nanehkaran, Y.A. (2020). Using deep transfer learning for image-based plant disease identification. *Computers and Electronics in Agriculture* 173. 105393, ISSN 0168–1699.

Chung, Chia-Lin & Huang, K.-J & Chen, Szu-Yu & Lai, M.-H & Chen, Y.-C & Kuo, Yan-Fu. (2016). Detecting Bakanae disease in rice seedlings by machine vision. *Computers and Electronics in Agriculture*. 121. 404–411. 10.1016/j.compag.2016.01.008.

Liu, Tao & Chen, Wen & Wu, Wei & Sun, Chengming & Guo, Wenshan & Zhu, Xinkai. (2016). Detection of aphids in wheat fields using a computer vision technique. *Biosystems Engineering*. 141. 82–93. 10.1016/j.biosystemseng.2015.11.005.

Prajwala, Tm, Pranathi, Alla, Kandiraju, Sai Ashritha, Chittaragi, B., Nagaratna & Koolagudi, Shashidhar. (2018). Tomato Leaf Disease Detection Using Convolutional Neural Networks. Eleventh International Conference on Contemporary Computing (IC3), 1–5. 10.1109/IC3.2018.8530532.

Shrivastava, Sourabh & Singh, Satish & Hooda, D S. (2017). Soybean plant foliar disease detection using image retrieval approaches. *Multimedia Tools and Applications*. 76. 10.1007/s11042–016–4191–7.

Szegedy, C., Vanhoucke, V., Ioffe, S., Shlens, J., & Wojna, Z. (2016). Rethinking the Inception Architecture for Computer Vision. *2016 IEEE Conference on Computer Vision and Pattern Recognition (CVPR)*, 2818–2826.

Tan, M. & Le, Q. (2019). EfficientNet: Rethinking Model Scaling for Convolutional Neural Networks. Proceedings of the 36th International Conference on Machine Learning, in Proceedings of Machine Learning Research 97:6105–6114 https://proceedings.mlr.press/v97/tan19a.html.

Too, E.C., Yujian, L., Njuki, S., & Yingchun, L. (2019). A comparative study of fine-tuning deep learning models for plant disease identification. *Comput. Electron. Agric.*, 161. 272–279.

Wang, G., Sun, Y., & Wang, J. (2017). Automatic Image-Based Plant Disease Severity Estimation Using Deep Learning. *Computational Intelligence and Neuroscience, 2017.*

11 Recognition of Facial Expressions in Infrared Images for Lie Detection with the Use of Support Vector Machines

*Rupali Dhabarde
Assistant Professor, Department of Technology,
Shivaji University, Maharashtra, India

D.V. Kodavade
Professor, DKTE society's Textile & Engineering Institute,
Maharashtra, India

Rashmi Deshmukh
Department of Technology, Shivaji University, Maharashtra,
India

Sheetal S. Zalte-Gaikwad
Department of Computer Science, Shivaji University,
Kolhapur, Maharashtra, India

*Corresponding author.

CONTENTS

DOI: 10.1201/9781003279044-11

163

11.1 INTRODUCTION

Different areas, such as finding interpersonal relationships, investigations, and terrorism, need to find facial expression, especially finding lie detection. Both ordinary and trained people fail to discriminate between liars and truth-tellers. These approaches used for detecting lies are inaccurate and time-consuming. Detecting and analyzing the facial expression of fear and lie is a challenging task. However, it is possible to do with computer vision and machine learning techniques. This study focuses on the classification of different facial expressions for deception detection using Infrared Images with the use of support vector machines (SVMs).

Researchers are showing their interest in developing software for Facial Expression Detection with the use of Deep Neural Network approaches. These approaches help in discovering parameters required for expression recognition. The dataset used for the purpose of classification is captured by the camera. These cameras are usually incorporated as an add-on with portable devices like phones and tablets, which might be reasonably-priced. Many researchers have already done work on FER from visual images, which are captured in the visible spectrum. In such cases, the facial expressions classification is a troublesome process due to consistent conditions like image shadow, reflection on the face, and obscurity. Besides, different things like background images, surroundings, and a wide range of issues exist along with the face.

Hence, it is hard to differentiate facial expressions. Infrared images use thermal distribution based on facial muscle mass and this helps in the classification of faces. Thermal imaging is used in various real-life applications. The main objective of this work is to study different techniques for recognition of facial expression for lie detection and to detect human faces using thermal imaging for feature extraction with the use of SVM.

11.1.2 DIFFERENT AVAILABLE MACHINE LEARNING ALGORITHMS

Machine Learning is the science of teaching machines on their own. Machines are made for doing repetitive things with more accuracy. Machines learning model helps to think machines. Machine learning models take data as input and use it to predict or classify.

The following are some of the commonly used machine learning algorithms:

1. **Linear Regression:** It is the most well-known algorithm in statistics and machine learning. Linear Regression does an analysis of predictions from the variables. It does predictions easily. With linear regression output values are found from the given input values. If the output of a single dependent variable is based on independent variables, then the equation is considered as simple linear regression. What's more, when there is more than one input variable, it is called multiple linear regression.

2. **Logistic Regression:** Logistic regression is a classification algorithm based on the function which is used at the core of the method, logistic function or sigmoid function. It's an S-shaped curve that is used to predict a binary outcome (1/0, Yes/No, True/False) given a set of independent variables. It can

also be thought of as a special case of linear regression when the outcome variable is categorical, where we are using the log of odds as a dependent variable. Also, it predicts the probability of occurrence of an event by fitting data to a logit function.

3. **K-Nearest Neighbors:** K-nearest neighbors (KNN) algorithm is used for performing the tasks of regression and classification. It works by finding the distances between the new data point added and the points already existing in the two separate classes. Whatever class got the highest votes, the new data point belongs to that class.

4. **Support Vector Machines:** It is a very popular supervised learning algorithm. It is mainly used in classification problems. An SVM will find a hyperplane or a boundary between the two classes of data that maximizes. There are other planes as well which can separate the two classes, but only the SVM hyperplane can maximize the margin between the classes.

5. **Decision Trees:** Decision tree algorithms are referred to as CART or Classification and Regression Trees. It is a flowchart like a tree structure, where each internal node denotes a test on an attribute, each branch represents an outcome of the test, and each leaf node (terminal node) holds a class label.

6. **Random Forest:** Random Forests are an ensemble learning technique that builds off of decision trees. Random forests involve creating multiple decision trees using bootstrapped datasets of the original data and randomly selecting a subset of variables at each step of the decision tree. And the model then selects the mode of all of the predictions of each decision tree (bagging).

7. **Naive Bayes:** It is a classification algorithm used for binary (two-class) and multiclass classification problems. It is used when the output variable is discrete. As the name specifies, this algorithm is entirely based on the Bayes theorem. Bayes' theorem says we can calculate the probability of a piece of data belonging to a given class if prior knowledge is given. These Machine learning algorithms are used to solve most data problems.

11.2   BACKGROUND AND LITERATURE REVIEW

The paper suggests that a facial expression is the noticeable appearance of the emotion, intellectual activity, intention, character, and psychopathology of an individual and assumes an open part in interpersonal relations. It has been read up for a significant stretch of time and acquired the advancement late many years ago. However, much headway has been made, perceiving facial expression with high precision continues to be troublesome because of the intricacy and varieties of expressions (Shan et al., 2005).

The paper demonstrates that for the most part, individuals can pass on goals and feelings through nonverbal ways such as motions, looks, and compulsory dialects. This framework can be a significantly valuable, nonverbal way for individuals to speak with one another. The important thing is the way the framework smoothly identifies or concentrates the look from the image. The framework is developing consideration since this could be generally utilized in numerous fields like lie location, clinical evaluation, and human PC interface. The Facial Action Coding System (FACS), which was proposed in 1978 by Paul Ekman, is a very famous look investigation instrument (Bhatt, 2014).

The author said that in everyday nuts and bolts, people regularly perceive emotions by trademark features, displayed as a piece of a look. For example, joy is obviously associated with a grin or a vertical development of the sides of the lips. Correspondingly, other emotions are described by different disfigurements regular to a specific articulation. Research into an automatic acknowledgment of looks resolves the issues encompassing the representation and classification of static or dynamic qualities of these deformations of face pigmentation (Chibelushi, 2003).

This paper presents and assesses the staggered arrangement structure for the facial expression. As per this study, the model consists of three main parts, i.e., face localization, feature extraction, and classification. By using multilevel classification, PCA and SVM results have been drawn (Drume, 2012).

The paper centers on another technique for face expression acknowledgment. Haar capacities are utilized for face, eyes, and mouth location; edge recognition strategy for extricating the eyes accurately and Bezier bends is applied to inexact the separated areas. Then, at that point, a bunch of distances for different face types is removed and it is filled in as preparing input for a multi-facet neural organization. The original component of this methodology comprises applying Bezier bends to proficiently extricate the distances between facial parts. The pre-grouping is finished utilizing the K-means algorithm. A two-layered feed-forward neural organization is then utilized as a grouping instrument for the info pictures. The consistency of the outcomes is exhibited by the middle worth. The exhibition accomplished here is 82%. The technique is prominent to treat circumstances when the eyes are shut. Solid enlightenment varieties influence the outcomes (Banu, 2012).

The author has given different approaches to the application of thermal imaging which may be referred to apprehend physiological signals to lead them to significance in social dealings. This method allows recognition of cognitive methods through involuntary expressions (Goulart et al, 2019).

The research targets are executing RVM for facial classification of static images. For testing, the Cohn-Kanade database is used. They reported 90.84% acknowledgment rates for RVM for six general articulations. The error rate on account of RVM (9.16%) is contrasted with that of the SVM (10.15%) classifier. The significant viewpoint is that in the case of RVM classifiers the quantity of importance vectors (156) is more modest than that of help vectors (276) of SVM. This procedure of classification requires less memory and processing time (Datcu, 2017).

Results show that this approach effectively perceives expression with a 93% acknowledgment rate. The outcomes propose that the method is more precise for facial expression.

The author has used an image-based static look acknowledgment strategy for facial emotion detection and recognition. This method contains a module of face detection that has three state-of-the-art detectors. A classification method, and convolutional neural networks (CNN), and images from the SFEW 2.0 dataset are being classified into seven main emotions automatically (Yu,2015).

The researcher introduced two plans for learning the group loads of the organization reactions: by limiting the log probability misfortune, and by limiting the hinge loss. This technique produces cutting-edge results on the FER dataset. The technique achieves 61.29% precision.

The paper has perceived looks from the pictures given in the JAFFE database. They have advanced a blend of three unique kinds of approach to execute: Scale Invariant Features Transform (SIFT), Gabor wavelets, and Discrete Cosine Transform (DCT). Some pre-handling steps have been applied prior to extricating the highlights. Support Vector Machine (SVM) with spiral premise part work is used for the characterization of looks. They assessed the outcomes on the JAFFE information base and designed experiments that ruined individual ward and individual free strategies. While executing this method, some passionate states were misclassified with others. Miserable was misclassified with dread and shock. Essentially, anger is misclassified with disdain and to a little degree with misery. Nonpartisan is the main articulation in the pictures that are not misclassified with some other articulation (Aziz et. al., 2015).

The paper uses a transferable belief model (TBM) and rule-based decision system for classifying human facial expressions. Basic features like eyes, mouth, and eyebrows are taken into consideration (Hammal et al, 2007).

11.3 METHODOLOGY

The proposed system uses the following different steps for lie detection with the SVM classifier. Figure 11.1 shows flow of proposed system architecture.

The proposed system architecture consists of different steps. First, the thermal image is taken as an input to the system then from that input detection of the face will be carried out. Once the face is detected, extraction of features is done. These sets of extracted features are forwarded to the SVM Classifier. The SVM classifier will classify the emotions.

11.3.1 IR IMAGE ACQUISITION

Picking a suitable data set of images is an important task as the image specification has a critical effect on the output of the detecting emotions. There are multiple available datasets. In this study, the dataset used is taken from the kaggle given by Ngô Khang. This dataset contains two directories, one for thermal images and the other for normal/visual images of which in this study we have used the thermal directory. There are a total of 2,538 thermal facial images of varying resolutions. Preprocessing is done on these images.

11.3.2 FACE DETECTION

In this stage, the detection of the face of the input image is performed. Face detection involves finding one or more faces from an input image; this finding of the face is done by locating the coordinate of the face. Localization is done with finding the extent of the face.

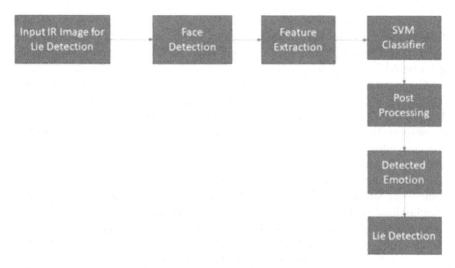

FIGURE 11.1   Flow of proposed system architecture.

11.3.3 FEATURE EXTRACTION

The process of converting the thermal image into a distinct and precise form is called feature extraction with the goal that the reference image and feature extraction can be compared. Feature extraction is completed on the Region of Interest that has been chosen.

11.3.4 CLASSIFICATION

SVM characterizes the input information into different classes by finding the maximum-margin hyperplanes, which is a line, plane, or hyperplane. This interaction intensifies the distance between the line and the nearest data elements. The SVM method can be summed up in three stages: The initial step is to find the hyperplanes in the feature space which can arrange input features. Since emotion detection for seven different expressions is a non-linear issue because of the high dimension of input highlights, mapping is carried out for each input sample to its representation in the feature space. Maximizing the margin and assessing the decision function both require the computation of the dot product in a high-dimensional space.

Machine learning algorithms help in predicting and classifying facial image data. Support Vector Machine (SVM) is used to resolve different classification or regression problems. It helps in solving linear and non-linear problems and operating well for different practical problems having a large number of features. The algorithm creates a line or a hyperplane which helps in distinguishing the data into classes. It helps to find an optimal boundary between the possible outputs. The basic objective of SVM is to find the nearest points that maximize the separation of the data points to their associated classes.

SVM is considered a good algorithm for image classification. As compared to the other classification and regression algorithms, the SVM method is totally distinctive due to SVM kernel functions. Kernel plays an important role in classification and is utilized to examine a few patterns in the given dataset. Kernels are very helpful in solving a non-linear problem by using a linear classifier. Further SVM algorithm uses kernel-trick for remodeling the data points and creating an effective decision boundary. Kernels assist us in managing high-dimensional data in a very effective way.

11.3.5 LIBRARIES AND PACKAGES OF SVM

To implement SVM on the selected dataset, open-source libraries for predictive analysis with machine learning libraries are used. These libraries and packages are developed in Python to perform classification, regression with SVM smoothly. Some of the functions of sklearn are SVC, NuSVC and LinearSVC. In LinearSVC, no value of the kernel is passed as it is specifically for linear classification.

The linear kernel can be used if data is linearly separable, since a linear kernel takes less training time as compared to other available kernel functions. It is preferred for text classification challenges as it performs smoothly for big datasets. If there is not much information regarding data that is not available Gaussian kernels give good results in such cases. While The Rbf kernel is a sort of Gaussian kernel that projects the high-dimensional data and afterward searches a linear separation for it and if the training data is normalized, Polynomial kernels can be used.

11.4 RESULTS AND DISCUSSIONS

The desired system is implemented using Python programming and four basic emotions are recognized: Happy, Sadness, Neutral, and Disgust.

We have used thermal facial images as input for the SVM classifier. Thermal face images are formed with the body heat pattern of the human being. As thermal IR sensors are used to capture the images, these images are independent of different lighting conditions. The thermal patterns of faces are generated from the pattern of superficial blood vessels under the skin. The vein and tissue structure of the face is unique for each person and, therefore, the IR images are also unique. Unique identification is made by finding the patterns of specific organs like eyes or parts used for biometric identification.

The infrared spectrum has four sub-bands, i.e., close to IR (NIR), short wave IR (SWIR), mediumwave IR (MWIR), and long-wave IR (LWIR). Most of the time the heat energy radiated from the IR spectrum is LWIR sub-band and after LWIR, MWIR does additionally radiate a massive quantity of heat-wave. As per the person's emotion, the temperature on the face increases or decreases. When the person is happy the temperature on the nose is decreases, while in sadness the chin temperature rises, and the forehead temperature decreases. While in disgust mood; the temperature of the forehead and near the tip of the eye increases. If the person is in neutral condition, then the nose temperature increases. With this, we have detected emotions using thermal images.

Dataset taken from Kaggle, contains two directories, one for thermal images and the other for normal/visual images of which we have used the thermal directory. A total of 2,538 thermal facial images with varying resolutions 353x470, 413x554, 209x282 are used as the source dataset for the experimentation. Image files are named in format IR_IR_ [EMOTION] _ [image id].jpg, we used this pattern and grouped the images in separate directories for each type of emotion. Images used for each

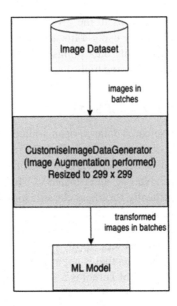

FIGURE 11.2   Person is honest (happy emotion).

category of emotion are anger: 280, disgust: 220, fear: 440, happy: 460, neutral: 398, sadness: 480, surprise: 260.

We have shown four different cases as follows:

Various moods of a person are detected from the thermal variations obtained on the image. Depending on the moods skin temperature varies due to blood circulation. Generally used RoI for emotion detection are the forehead, nasal tip, chin, cheek, etc. The first case Figure 11.2 shows the happy mood of the person, hence we can say that he is not lying, so considered as honest. In the following figures, the red spot shows the higher temperature of the face detected by the thermal camera. Different facial expressions are categorized into different emotions from the color pattern obtained on the image. In the happy mood generally, the upper cheek part of the face shows the higher temperature as shown in Figure 11.2.

I. Emotion: **happy**

Similarly, Figure 11.3 shows sadness, as thermal temperature is more at the forehead area captured by the IR camera, while Figure 11.4 shows a neutral person.

II. Emotion: **Sadness**
III. Emotion: **Neutral**

Figure 11.5 shows the high temperature of the forehead and near the tip of the eye, so, this person is considered to be lying.

IV. Emotion: **Disgust**

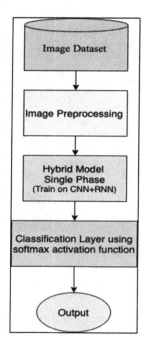

FIGURE 11.3 Person is honest (sad emotion).

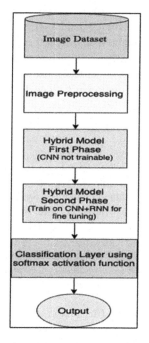

FIGURE 11.4 Person is honest (neutral emotion).

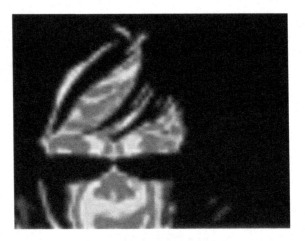

FIGURE 11.5   Person is lying (disgust emotion).

11.5   CONCLUSION AND FUTURE WORK

Face detection technology has various applications in video surveillance systems like maintaining security, detection of any object, etc. The article uses the SVM classifier technique for classifying the emotions of a face. Four basic emotions like neutral, disgust, happy and sad are recognized by the classifier. We have used this classifier for the detection of lies. As SVM algorithm performs well for pattern classification and regression-based methods; used for image classification. Experimentation shows that SVMs give higher search accuracy and better image segmentation systems than traditional query refinement methods and neural networks. This approach is more suitable for dynamic and interactive approaches used for the recognition of facial expressions. Its implementation can be extended in a direct manner to incorporate various techniques of feature extraction as well as different classifiers.

REFERENCES

Aziz M.et. al. (2015), Facial Expression Recognition using Multiple Feature Sets. IEEE. DOI: 10.1109/ICITCS.2015.7292981.

Banu, S. M. et. al. (2012). A Novel Approach for Face Expressions Recognition. IEEE 10th Jubilee International Symposium on Intelligent Systems and Informatics.

Bhatt, M., Drashti, H., Rathod, M., Kirit, R., Agravat, M., & Shardul, J. (2014). A Study of Local Binary Pattern Method for Facial Expression Detection. In 2014 arXiv preprint arXiv: 1405.6130.

Chibelushi, C. C., & Bourel, F. (2003). Facial expression recognition: A brief tutorial overview. In *2013 CVonline: On-Line Compendium of Computer Vision*, 9, available at https://homepages.inf.ed.ac.uk/rbf/CVonline/LOCAL_COPIES/CHIBELUSHI1/CCC_FB_FacExprRecCVonline.pdf

Datcu, D. and Rothkrantz, J. M. (n.d.). Facial Expression Recognition with Relevance Vector Machines. Delft University of Technology Man-Machine Interaction Group

Drume, D. and Jalal, A. S. (2012). A Multi-level Classification Approach for Facial Emotion Recognition, IEEE International Conference on Computational Intelligence and Computing Research.

Goulart, C., Valadão, C., Delisle-Rodriguez, D., Caldeira, E. & Bastos, T. (2019). Emotion analysis in children through facial emissivity of infrared thermal imaging. *PLoS ONE* 14, e0212928.

Hammal, J., Covreur, L., Caplier, A. & Rombout, M. (2007). Facial expression classification: An approach based on the fusion of facial deformations using the transferable belief model. *International Journal of Approximate Reasoning* 46, 542–567.

Shan, C., Gong, S., & McOwan, P. W. (2005, September). Robust facial expression recognition using local binary patterns. ICIP 2005. IEEE International Conference on Image Processing, 2, II–370.

Zhiding, Yu et al. (2015). Image based Static Facial Expression Recognition with Multiple Deep Network Learning. Proceedings of the 2015 ACM on International Conference on Multimodal Interaction https://doi.org/10.1145/2818346.2830595

12 Support Vector Machines for the Classification of Remote Sensing Images
A Review

Parijata Majumdar
Techno College of Engineering Agartala

*Suchanda Dey
Techno College of Engineering Agartala

Sourav Bardhan
Techno College of Engineering Agartala

Sanjoy Mitra
Tripura Institute of Technology, Narsingarh

*Corresponding author.

CONTENTS

12.1   INTRODUCTION

The increasing number of channels and the increase in spatial resolution have prompted the development of remote sensing processing technologies. Various processing and application schemes with image algorithms and recognition methods are used in these sectors. The objective of the work given here is to segment land and create a map of multispectral image categorization. Our goal is to create a data from fusion. After that, a categorization method based on learning is used to combine the data. Data collected from a distance is used in a variety of applications. To turn data into useful information, an image categorization procedure is usually started. Image

DOI: 10.1201/9781003279044-12

categorization is unfortunately not a simple job. Remote sensing data classification is very tough due to the fact that most supervised learning algorithms require a large number of training samples, while defining and acquiring reference data is frequently a challenge as mentioned in Chi et al. (2008). In many situations, including remote sensing, various parametric and non-parametric categorization algorithms have been developed and used.

Earlier research articles pointed to current advances in methods for a specific sort of image processing, like hyperspectral images as stated by Plaza et al. (2009). This paper's review is based on algorithmic considerations rather than image attributes. We concentrate on the use of support vector machines (SVMs) in sensing of remote images.

Desire to conduct this research stems from a variety of reasons. First, while SVMs are not as well known in the remote sensing world as other classifiers (e.g., neural network variations, decision trees), they can match, if not outperform, existing approaches. Gains appear to be well-suited to distant sensing tasks, where only a small amount of data for referencing is frequently available. Third, while the method is not extensively used, it has been noticed that there has been a large growth in SVM work on sensing of remote images problems in recent years, indicating that this survey is relevant and suitable to carry out.

This survey concentrates on newer research articles published in eight major remote sensing journals: Photogrammetric Engineering & Remote Sensing, Remote Sensing of Environment, IEEE Transactions on Geoscience and Remote Sensing, ISPRS IEEE Geoscience and Remote Sensing Letters, Journal of Photogrammetry and Remote Sensing, and IEEE Transactions on Geoscience and Remote Sensing, to name a few. Additional sources provided a very limited number of research papers which is related to the theme topic and thus incorporated in this survey. The papers chosen, cover a wide range of topics, including pixel sizes ranging from sub-meter to several kilometers, spectral resolution ranging from single to hundreds of bands, and methods ranging from maximum likelihood classifiers to neural networks for comparison purpose, and pixel sizes ranging from sub-meter to several kilometers. We will go over the fundamentals of SVM methodology first to make sure we are all on the same page before diving into specific works. Relevant publications are described, and we can draw inferences and make recommendations for future research based on the juxtaposition of general trends.

12.2   MOTIVATION OF THE REVIEW

Support Vector Machine (SVM) and image analysis techniques can be used to locate hotspots in agricultural areas using NOAA/AVHRR satellite data. One of the major advantages of using satellite data is that it is free and has good temporal resolution; unfortunately, it has poor spatial resolution (i.e., approximately 1.1 km). As a result, it is vital to conduct research employing effective optimization approaches and image analysis techniques to overcome these weaknesses, as well as to use satellite remote images for effective hotspot detection and monitoring. For hotspot detection, SVM and multi-threshold algorithms are being researched. To eliminate cloud coverage from land coverage a multi-thresholding method can be used. This algorithm can be used to highlight hotspots or fire locations in suspicious areas. SVM has an advantage over multi-thresholding in that it can learn patterns from examples, allowing it to optimize performance by removing spurious points that the threshold technique highlights. Either approach can be used alone or in combination, depending on the size of the image. The RBF (Radial Basis Function) kernel can be used to train the following sets of inputs: brightness temperature. The Global Environment Monitoring Index (GEMI) and the Normalized Difference Vegetation Index (NDVI) are two different forms of vegetation indices. These indexes can be used to categorize an image that emphasizes the hotspot and non-hotspot pixels. The performance of the SVM may be compared to that of neural networks, and it appears that the SVM detects hotspots more accurately (over 91 percent classification accuracy) and with a lower false alarm rate. The findings are in good agreement with observations obtained on the ground at hotspots. This type of research will be extremely important in the development of a hotspot monitoring system based on brightness temperature using operational satellite data in the near future. Figure 12.1 describes the role of SVM to classify O/P Satellite Remote Sensing Images.

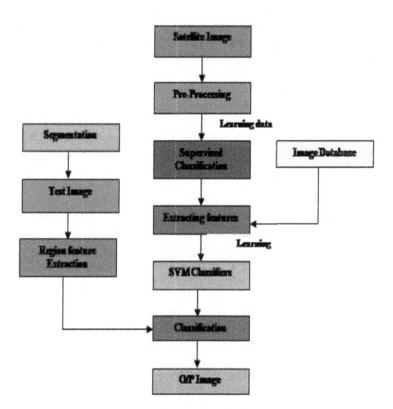

FIGURE 12.1   Role of SVM to classify O/P satellite remote sensing images.

12.3 RELEVANT LITERATURE REVIEW

SVMs are particularly appealing in the field of remote sensing because they require less training data and provide higher classification accuracy than earlier methods. Each image instance was treated to the same random permutation when the pixels were originally jumbled, and the accuracy was almost identical. Despite this, when "vandalism" (the removal of prior knowledge) was conducted, SVM beat even the best neural networks. Even if the data were distributed regularly, with a dimensionality equal to the number of spectral bands, the belief that the distribution might be described by a bell curve persists. Because the concentration has grown, the Gaussian function is no longer valid. Data tends to be in the tails in higher-dimensional space. Several studies provide a thorough explanation of SVM kernels and how they work. None of them, however, go into detail on the expanding number of new SVM variants being proposed. Support vector machines (SVMs) have lately become popular in remote sensing. For this review, we found 108 papers that were relevant, with more than half of them published within the last 2.5 years. This growing trend is expected to continue, making now a critical time to evaluate current work. The implementation algorithm in a geological classification was an SVM, which showed greater robustness to small data samples. Blanzieri and Melgani (2008) looked at how to create localized SVMs techniques based on k-nearest neighbor adaptation on a local scale. Significant increases were seen, especially when nonlinear kernel functions were used. To overcome the problem of kernel predetermination, Tuia and Camps-Valls (2009) suggested a regularization strategy for finding kernel structure using unlabeled samples. Camps-Valls et al. (2010) enhanced the Hilbert–Schmid technique for proving kernel independence in a variety of picture kinds. Ghoggali et al. (2009) used a limited number of training data to test the combination of a genetic algorithm (GA) with a support vector machine (SVM) for remote sensing categorization. The recommended SVM technique outperformed the other classification models in terms of resilience and efficacy when non-fully reliable training samples were used. Huang and Zhang (2010) compared multi-SVM methods to typical vector stacking techniques in high-resolution urban mapping. By integrating identifiable and unlabeled pixels, Gomez-Chova et al. (2010) advocated employing clustering and the mean map kernel to increase classification reliability and accuracy. They tested their cloud classification system using data from Envisat's Medium Resolution Imaging Spectrometer (MERIS). They discovered that their technique works best when sample selection bias exists, i.e., training and test data have different distributions. The genetically optimized SVM using the support vector counts as a criterion to achieve the best results for both simulated and real-world AVIRIS hyperspectral data. Mathur and Foody (2008) investigated the performance of SVMs in non-binary classification tasks. Their method outperformed the change vector analysis CVA-based method using the expectation-maximization technique, but it took much longer to compute due to the model-selection process they utilized to identify the optimum structure for their model. Mukhopadhyay and Maulik (2009) utilized a multi-objective fuzzy clustering technique with an SVM for unsupervised classification. They discovered that their method outperformed traditional SVMs and was less affected by the size of the training sample. Using in situ hyperspectral data, Sun et al. (2009) employed SVMs to estimate chlorophyll concentration in Lake Taihu. They used an iterative optimization technique to get the best three-wavelength factor, which they then fed into an SVM to estimate chlorophyll content. Traditional linear regression models were proven to be less reliable than their method. Knudby et al. (2010) studied reef fish richness, diversity, and biomass using Ikonos images and predictive modelling. SVMs fared nearly as well as the top-ranked ensemble algorithms when compared to five other techniques. Clevers et al. (2007) used an SVM-based band shaving strategy to study how to reduce dimensionality in hyperspectral datasets. Grassland biomass estimation was the application domain, and three bands were determined to be sufficient for field investigations. The worldwide ocean primary production was calculated using an SVM-based model. It was discovered to be more productive. VGPM (vertically generalized production model) approach is more accurate because of its capacity to detect nonlinear relationships between the primary production of the ocean, and other factors. The scant nature of the challenge made it very tough. According to Xie et al. (2008), SVR outperformed linear regression and backpropagation neural networks when calculating moisture transport in maritime environments using MISR. SVMs outperformed the k-NN method and Gaussian maximum likelihood classification. They claimed that incorporating LiDAR variables increased classification performance in general, with the first return data being the most important component. To assess tree species classification, Heikkinen et al. (2010) employed an aerial four-band sensor system and a simulated optical radiation model.

12.4   CONCLUSION AND FUTURE WORK

This research emphasized the importance of SVM-based work in remote sensing. The majority of the findings suggest that the theoretical formulation and motivation for SVMs are supported by empirical evidence. SVM's most important attribute is its ability to generalize successfully from a limited amount and/or quality of training data. In comparison to other methodologies, such as backpropagation neural nets, SVMs can achieve comparable accuracy with a much smaller training sample size. This is comparable to the "support vector" notion, in which the classifier's hyperplane is formed by only a few data points. In remotely sensed data, outlier effects are common, and SVMs are not designed to deal with the underlying problem of noisy data. Impurities are caused by measurement mistakes, caused by the limited precision of image acquisition systems, as well as atmospheric and topographic distortions. The quality of both training and test patterns is crucial in the construction (training) and evaluation of automatic classification, identification, and detection systems. An SVM classifier's performance can be severely harmed by a small number of mislabeled data. RVMs are a Bayesian treatment option for SVMs that offer various benefits such as probabilistic predictions, automatic parameter estimation, and flexible kernel functions. According to the authors, the novel approach minimizes hyperparameter sensitivity, allowing non-Mercer kernels to be used. RVMs also allow for fuzzy (or sub-pixel) data categorization, which yields a probabilistic outcome. In conclusion, SVM classifiers have proven to be a fairly reliable methodology in the intelligent processing of data obtained via remote sensing, as they are characterized by self-adaptability, rapid learning speed, and small training size needs. SVMs have been shown to beat most other algorithms in both real-world and simulated contexts, which is a key driver and promise for future developments.

REFERENCES

Blanzieri, E., Melgani, F., 2008. Nearest neighbor classification of remote sensing images with the maximal margin principle. *IEEE Transactions on Geoscience and Remote Sensing* 46 (6), 1804–1811.

Camps-Valls, G., Mooij, J., Scholkopf, B., 2010. Remote sensing feature selection by kernel dependence measures. *IEEE Geoscience and Remote Sensing Letters* 7 (3), 587–591.

Chi, M., Feng, R., Bruzzone, L., 2008. Classification of hyperspectral remote-sensing data with primal SVM for small-sized training dataset problem. *Advances in Space Research* 41 (11).

Clevers, J.G.P.W., van der Heijden, G.W.A.M., Verzakov, S., Schaepman, M.E., 2007. Estimating grassland biomass using SVM band shaving of hyperspectral. *Data Photogrammetric Engineering & Remote Sensing* 73 (10), 1141–1148.4.

Ghoggali, N., Melgani, F. and Bazi, Y., 2009. A multiobjective genetic SVM approach for classification problems with limited training samples. *IEEE Transactions on Geoscience and Remote Sensing*, 47(6), 1707–1718.

Gomez-Chova, L., Camps-Valls, G., Bruzzone, L., Calpe-Maravilla, J., 2010. Mean map kernel methods for semisupervised cloud classification. *IEEE Transactions on Geoscience and Remote Sensing* 48 (1), 207–220.4.

Heikkinen, V., Tokola, T., Parkkinen, J., Korpela, I., Jaaskelainen, T., 2010. Simulated multispectral imagery for tree species classification using support vector machines. *IEEE Transactions on Geoscience and Remote Sensing* 48 (3), 1355–1364.

Huang, X., Zhang, L., 2010. Comparison of vector stacking, multi-SVMs fuzzy output, and multi-SVMs voting methods for multiscale VHR urban mapping. *IEEE Geoscience and Remote Sensing Letters* 7 (2), 261–265.

Knudby, A., LeDrew, E., Brenning, A., 2010. Predictive mapping of reef fish species richness, diversity and biomass in Zanzibar using IKONOS imagery and machine-learning techniques. *Remote Sensing of Environment* 114 (6), 1230–1241.

Mathur, A., Foody, G.M., 2008. Multiclass and binary SVM classification: implications for training and classification users. *IEEE Geoscience and Remote Sensing Letters* 5 (2), 241–245.

Mukhopadhyay, A., Maulik, U., 2009. Unsupervised pixel classification in satellite imagery using multiobjective fuzzy clustering combined with SVM classifier. *IEEE Transactions on Geoscience and Remote Sensing* 47 (4), 1132–1138.

Plaza, A., Benediktsson, J.A., Boardman, J.W., Brazile, J., Bruzzone, L., Camps-valls, G., Chanussot, J., Fauvel, M., Gamba, P., Gualtieri, A., Marconcini, M., Tilton, J.C., TriannI, G., 2009. Recent advances in techniques for hyperspectral image processing. *Remote Sensing of Environment* 113 (1), S110–S122.

Sun, D., Li, Y., Wang, Q., 2009. A unified model for remotely estimating chlorophyll a in Lake Taihu, China, based on SVM and in situ hyperspectral data. *IEEE Transactions on Geoscience and Remote Sensing* 47 (8), 2957–2965.

Tuia, D., Camps-Valls, G., 2009. Semisupervised remote sensing image classification with cluster kernels. *IEEE Geoscience and Remote Sensing Letters* 6 (2), 224–228.

Xie, X., Liu, T., Tang, B., 2008. Spacebased estimation of moisture transport in marine atmosphere using support vector regression. *Remote Sensing of Environment* 112 (4), 1846–1855.

13 A Study on Data Cleaning of Hydrocarbon Resources under Deep Sea Water Using Imputation Technique-Based Data Science Approaches

Dipjyoti Deb
Department of Electronics & Communication Engineering,
Techno College of Engineering Agartala, West Tripura, India

CONTENTS

DOI: 10.1201/9781003279044-13

13.1 INTRODUCTION

A hydrocarbon is an organic compound that is made up of hydrogen and oxygen and can be found in all types of crude oil, petroleum, and natural gases. Hydrogen and carbon are present in different trees and plants on the surface. Unfortunately, the surface resource of hydrocarbons gets reduced every day due to the rapid increase of industrialization [1]. Deep oceans can also be used as a source of hydrocarbons. Millions of years ago, microorganisms deposited and buried in the sediments over a period of time, and now by achieving sufficient pressure and temperature, the hydrocarbons are released. This hydrocarbon, in the form of gas, leaks and spreads in the ocean, posing a threat to sea life [2] [3].This undersea fossil fuel can be a giant resource of future energy and can become a factor in the development of the blue economy. Table 13.1 shows hydrocarbon gas liquids and their uses, products, and various sectors where they are used. Figure 13.1 depicts the daily hydrocarbon-based gas liquids consumption in the United States of America.

13.2 LITERATURE SURVEY

The primary task is data collection, which is to identify the hydrocarbon resource areas. Data mining has significantly advanced analytical tools that can be used in the

TABLE 13.1
Different Basic Compounds of Hydrocarbon and Their Uses and Products

Hydrocarbon Gas Liquids (HGL)	Uses	End-Use products	End-Use sectors
Ethane	Petrochemical feedstock for ethylene production; power generation	Plastics; anti-freeze; detergents	Industrial
Propane	Fuel for space heating, water heating, cooking, drying, and transportation; petrochemical feedstock	Fuel for heating, cooking, and drying; plastics	Industrial, residential, commercial, and transportation
	Fuel for space heating, water heating, cooking, drying, and transportation; petrochemical feedstock		
Butanes Natural gasoline (pentane plus)	Petrochemical and petroleum refinery feedstock; motor gasoline blending	Motor gasoline; plastics, synthetic rubber	Industrial and transportation
Refinery olefins (ethylene, propylene, normal butylenes and isobutylene)	Petrochemical feedstock; additive or motor gasoline; diluents for heavy crude oil Intermediate feedstock in petrochemical industry	Motor gasoline; ethanol denaturant; solvents	Industrial and transportation
		Plastics; artificial rubber; paints and solvents; resins	Industrial

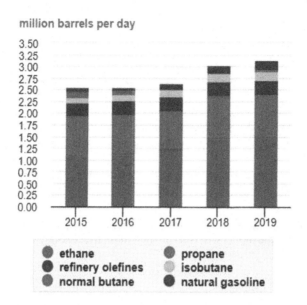

FIGURE 13.1 U.S hydrocarbon gas liquids consumption by type 2015–2019.

prediction of resources [4]. Data to be predicted can be collected using a geological survey, a seismological survey, or a magneto-graphic survey. Advancement in the area of remote sensing accelerates the data collection process from various difficult terrains as well [5]. In the last decade, satellite imagery has also received significant attention. There are five phases of the data cleansing process, viz., data analysis, definition of transformation workflow and mapping rules, verification, transformation, and lastly, backflow of cleaned data [6].

Sapna Devi et al. in their paper show a set of criteria for data quality, namely accuracy, integrity, completeness, validity, consistency, uniformity, density, and uniqueness [7]. S. Lakshmi et al. describe the method for handling quantitative and categorical datasets [8].

There are a number of imputation techniques that are available for data cleansing. K. Manimekalai et al. show a single imputation-based knowledge discovery from database (KDD) method [9]. Qinbao Song et al. [10] describe Bayesian estimation of PCA (BPCA) and hybridization of the KNN and BPCA methods for the data cleaning process. Another method is proposed by Vaishali Chandrakant Wangikar et al., where the author uses fuzzy match-based online data cleaning and clustering and association-based algorithm [11].

13.3 OVERVIEW OF DATA SCIENCE

Data science uses scientific methods, processes, algorithms, and systems to extract knowledge and insights from structured and unstructured data and apply knowledge and actionable insights from data across a broad range of application domains.

13.3.1   DATA SCIENCE LIFE CYCLE PHASES

Data collection: It is the process of collection of data of particular feature of a dataset.

Data preparation: It is the method of data cleaning or data wrangling to process further.

Exploratory data analysis: It has an important role at the stage of summarization of clean data that helps in identifying the structure, outliers and pattern of the data.

Data modelling: This process chooses the appropriate type of the model that the problem is based on: classification, regression, or clustering, and many more.

Model evaluation: It helps to choose the best model according to the dataset depending on how the model will work in the future. Model evaluation mainly divided in two sections:

1 Classification evaluation
2 Regression evaluation

Model deployment: It is the application of the model for prediction using a new data.
This chapter deals with the data cleaning which is the most important part of data preparation.

13.4   DATA CLEANING OR DATA WRANGLING PROCESSES

Data cleaning which is the part of data preparation primarily consists of four possibilities:

Removing unwanted observations: Removes duplicate observations and irrelevant data.

Missing data handling: Missing data of any feature cannot be neglected as there are several examples that do not support missing data. But for a feature, if 80% of the data is missing, then that feature column can be ignored. In the case of less than 80% missing blocks, they need to be filled up using several techniques. This chapter shows three methods: mean, median, and KNN algorithms to fill up missing data fields.

Structural error solving: It consists of unknown naming or conventions or incorrect capitalizations etc.

Outliers management: It is the case of abnormal distance from other values in a random sample from a population. Such as presence of categorical data in a numeric feature's column.

13.5 METHODOLOGY

The same method of data cleansing may not be suitable for every domain and also not for every feature. In this chapter, an imputation technique-based approach is applied to clean the dataset and to achieve structured data. We have taken the "Polycyclic Aromatic Hydrocarbon (PAH) Samples dataset (2011 to 2021)" from the California Department of Fish and Wildlife [https://catalog.data.gov/dataset/polycyclic-aromatic-hydrocarbon-samples-2011-to-2021-ospr-ds714-1fc7a].

In this approach, we first look into the relationship between the different PAHs. In nature, based on different environmental parameters, the abundance of PAHs varies and thereby creates different impact levels. While collecting this data, we might find some data missing because of this variation. However, for our machine learning, we might need to impute some of this missing data. As we have mentioned above, we look into it in steps. For the implementation, we are using Python, Pandas, Numpy, and SKlearn.

Before we start the analysis, here are the assumptions and properties of our current dataset:

- The readings are taken at random points in time and do not have any time series continuation.
- Currently, only a limited number of readings are available, which is around 300.
- The PAH quantities are more likely to be found in similar amounts in similar climatic conditions.
- All the numeric PAH values that are recorded are taken by instruments that are properly calibrated before taking them, but they still may contain a few erroneous records.

Our objective of our current paper is to find which traditional imputation techniques can be applied to PAH data and then select between "mean," "median," "knn," "decision tree imputation," and "linear regression imputation" algorithms for individual features. However, it can be extended to evaluate any new algorithm that best fits the imputation approach with the current data.

13.5.1 Imputation-Based Algorithms

13.5.1.1 Imputation Using Mean

The "Mean" algorithm fills the missing data in the feature with the mean value of the remaining dataset. This is helpful in cases where the values are pretty much concentrated towards the center of the frequency distribution.

13.5.1.2 Imputation Using Median

The missing values in the "Median" algorithm are filled with the feature's 50th percentile value. This method is appropriate for columns with fewer missing values and possibly outliers.

13.5.1.3   Imputation Using KNN

The "KNN" algorithm, also known as the k-nearest neighbor algorithm, is based on the Euclidean distance between a node and its k-nearest neighbors. According to the dataset properties mentioned above, PAH traces can be grouped by climatic conditions, so the KNN algorithm can be useful in such scenarios. The details of the KNN algorithm with pseudo-code are given in the subsequent parts of this chapter.

13.5.1.4   Imputation Using Decision Tree

In the case of datasets with more categorical values, the "decision tree" imputation method can be applied to impute the missing values. In our case, however, we want to impute numerical values so that we can rule out decision trees for imputation.

Also, we can leave linear regression in our dataset because, in spite of a few correlated features, we need to consider other important environmental factors for our imputations, even with less correlation.

13.5.2   ROOT MEAN SQUARE ERROR (RMSE)

In machine learning, the performance of the model using a single number is very helpful during training, cross-validation, or after deployment. RMSE is very helpful in this case. It is used to find the quality of the prediction. It finds how far the prediction falls from true or actual value using Euclidean distance. So, as the RMSE value decreases, the error between the predicted and actual value will also decrease.

$$RMSE = \sqrt{\frac{\sum_{i=1}^{N}\left\|y(i)-\hat{y}(i)\right\|^2}{N}} \tag{13.1}$$

Where,

   N = *number of data points*
   $y(i)$ = *true or actual value of ith sample*
   $\hat{y}(i)$ = *corresponding prediction of ith sample*

13.5.3   CORRELATION

It is a statistical measurement that determines the relationship between variables. If two variables are correlated, then the value of an item of one variable can be predicted with the help of another variable.

Positive correlation: When one variable increases, and other one is also increases.
No correlation: When variables does not have relation.
Negative correlation: When one variable increases, other one is decreases.

   In other words, correlation coefficient shows the strength of the relationship of two variables.

$$r_{xy} = \frac{\sum (x_i - \bar{x})(y_i - \bar{y})}{\sqrt{\sum (x_i - \bar{x})^2 \sum (y_i - \bar{y})^2}} \qquad (13.2)$$

Where,

r_{xy} = *correlation coefficient*
x_i = *values of x variables in the sample*
$\bar{x}$ = *mean of the values of the x variable*
y_i = *values of y variables in the sample*
$\bar{y}$ = *mean of the values of the y variable*

13.5.4   K-NEAREST NEIGHBOR ALGORITHM

It is a supervised learning method. It is used to solve classification and regression type problems.

Pseudo-Code for KNN Algorithm

Take a data set and distribute all the data points to a number of classes. Now, for an unknown data point, this algorithm will predict the class in which it should belong.

Step 1: Load the data set.
Step 2: Initialize the total number of neighbor points in the data set. Let K's neighbor point out.

To get the predicted class for the unknown point, iterate from 1 to the total number of training data points.

(a) Calculate the Euclidean distance between the unknown point and the training data points and create an array.
(b) Sort the calculated distances in ascending order.
(c) Consider the top K rows from the sorted array.
(d) Check the most frequent class for these K rows.
(e) Unknown data points will belong to the most frequently occurring class. Return to the predicted class.

Firstly, to handle missing data, starting by data profiling to understand the nuances of the data. It will give an aggregated matrix of meta information for each column, viz. the categorical or numerical columns, min-max, percentiles, number of unique, null proportions etc. This knowledge of the data will help in targeting the features for imputation. Once the data profiling is done, start to identify columns with more than 80% missing values. These columns can be ignored by feature engineering.

Secondly, look into the correlation between the PAH columns, latitude-longitude, and other measured columns using the correlation matrix. Based on the correlation coefficient, they create sets of features which can be used to impute together.

	X	Y	OBJECTID	Latitude	Longitude	Mass_Vol_Extrc	Dry_Samp_Mass	Moisture	Naph	F2_MeNaph	F1_MeNaph	F2_6_MeNaph	F2_3_5_MeNaph	C1_Na
X	1.000000	-0.939042	-0.268914	-0.941907	1.000000	0.080896	0.022813	-0.092368	-0.075749	-0.186445	-0.207012	-0.172268	-0.157825	-0.2019
Y	-0.939042	1.000000	0.152352	0.999924	-0.939042	-0.071729	0.025777	0.086772	0.055143	0.159604	0.156743	0.155723	0.132526	0.1494
OBJECTID	-0.268914	0.152352	1.000000	0.163130	-0.268914	0.020725	-0.140675	-0.002762	0.281282	0.232168	0.317013	0.253105	0.394853	0.2857
Latitude	-0.941907	0.999924	0.163130	1.000000	-0.941907	-0.072311	0.014707	0.093246	0.059223	0.164057	0.161404	0.156358	0.135610	0.1519
Longitude	1.000000	-0.939042	-0.268914	-0.941907	1.000000	0.080896	0.022813	-0.092368	-0.075749	-0.186445	-0.207012	-0.172268	-0.157825	-0.2019
Mass__Vol_Extrc	0.080896	-0.071729	0.020725	-0.072311	0.080896	1.000000	0.344225	-0.037925	-0.070968	-0.054264	-0.048993	-0.064291	-0.097634	-0.0550
Dry_Samp_Mass	0.022813	0.025777	-0.140675	0.014707	0.022813	0.344225	1.000000	-0.996091	-0.437506	-0.015519	-0.153681	-0.051263	-0.353811	-0.0939
Moisture	-0.092368	0.086772	-0.002762	0.093246	-0.092368	-0.037925	-0.996091	1.000000	0.241011	0.088885	0.095256	-0.007705	0.026194	0.0970
Naph	-0.075749	0.055143	0.281282	0.059223	-0.075749	-0.070968	-0.437506	0.241011	1.000000	0.669267	0.732375	0.691705	0.561813	0.6556
F2_MeNaph	-0.186445	0.159604	0.232168	0.164057	-0.186445	-0.054264	-0.015519	0.088885	0.669267	1.000000	0.927365	0.757251	0.474381	0.9829
F1_MeNaph	-0.207012	0.156743	0.317013	0.161404	-0.207012	-0.048993	-0.153681	0.095256	0.732375	0.927365	1.000000	0.788342	0.572010	0.9180
F2_6_MeNaph	-0.172268	0.155723	0.253105	0.156358	-0.172268	-0.064291	-0.051263	-0.007705	0.691705	0.757251	0.788342	1.000000	0.860650	0.8259
F2_3_5_MeNaph	-0.157825	0.132526	0.394853	0.135610	-0.157825	-0.097634	-0.353811	0.026194	0.561813	0.474381	0.572010	0.860650	1.000000	0.5881
C1_Naph	-0.201997	0.149498	0.285732	0.151957	-0.201997	-0.055013	-0.093968	0.097097	0.655639	0.982964	0.918019	0.825962	0.588186	1.0000
C2 Naph	-0.151874	0.126371	0.225946	0.129996	-0.151874	-0.046137	-0.093002	0.002387	0.470830	0.569925	0.558139	0.915448	0.977034	0.7706

✓ 0s   completed at 10:15 AM

FIGURE 13.2   Correlation between features (Program result).

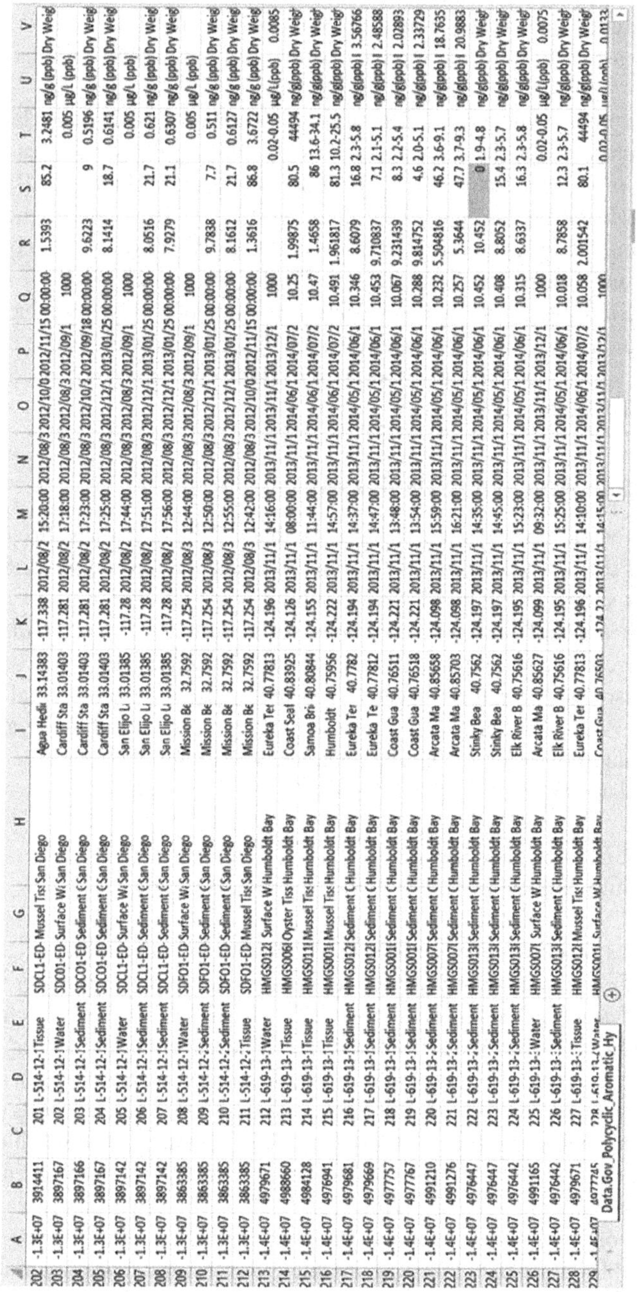

FIGURE 13.3 Outliers in the feature (Program result).

For the current dataset, consider all features with a correlation factor of > 0 in a single matrix to impute. There are multiple algorithms for imputations. So, to identify which algorithm will give the best results, pick a feature and create a train-test data set. In this current data set, the "Naph_d8" column (column-BR of the original data set) was chosen to find the best approach which has the highest correlation factor, Figure 13.2.

For outliers management, which is also an essential part of data cleaning, the "Moisture" feature (column-S) is chosen. It can be easily identified that for object id-222, moisture is zero Figure 13.3. This data belongs to "Humboldt Bay," which can be verified from the "Region" feature (column-H), which is simply not possible because Humboldt Bay is a coastal area, where moisture is very high compared to other regions and can not be zero. To remove this outlier, consider the maximum moisture value among all the moisture values in Humboldt Bay to replace the zero value. This is a threshold-based imputation technique. As moisture is one of the most vital features, mean, median, or mode-based algorithms may not be appropriate because they will impact other data.

13.6   RESULTS AND DISCUSSIONS

First, neglect feature columns that have more than 80% null values. So, out of 300 rows from this dataset, we drop those feature columns that have more than 240 rows of null data, as shown in Figure 13.4.

Now create a set of the most correlated features to the target feature "Naph_d8." In the current dataset, it mostly correlates to "Biphenyl_d10." Apart from these features, we also add "moisture" because it is an important environmental parameter with regard to PAHs. With this data frame, first drop all rows with null values in the "Naph_d8" column. For the remaining data, take the last 30 values of the target feature from the data frame.

(a)   Store it in a new data frame.
(b)   Force fully nullify the values in the original data frame.

These 30 values will be used as a test set to verify our imputed values.

To start with, we created three different imputers, viz., Simple-Imputer with Mean strategy, Simple-Imputer with Median strategy, and KNN-Imputer, which are available in the SKlearn library. With the data being fed to the imputers, we run the "transform" function to create the "Naph_d8_imputed" column with the imputed values in the 30 rows where we forcefully nullified the values. Then create a results data frame with the original column value and the imputed column value. The result data frame can be passed through a "Root Mean Square Error" function (RMSE) to find the error deviation of the imputed value from the actual value. The RMSE is a measure of the squared mean of the residual errors of the actual and predicted. This means that the lower the RMSE value, the closer the imputed values would be to actual readings. This would be the loss function for the algorithms.

C2_Fluor	241
F2_3_5_MeNaph	245
C4_Phen_Ant	246
C3_DBT	248
C3_Fluor	248
D_a_h_Ant	249
F4_MeDBT	253
F2_MeFluorant	253
F9_MePhen	256
C2_Fluorant_Pyr	256
F3_6_DiMePh	257
F3_MePhen	259
F2_MePhen	259
F1_MeFluor	266
Ret	267
C3_Fluorant_Pyr	268
Dec	268
trans_Dec	269
F2_MeAnt	270
C1_Dec	271
B_b_NTP	274
C4_Fluorant_Pyr	275
Fluor_d10	276
F2_MeDBT	276
F1_MeDBT	276
Terph_d14	276
Fluorant_d10	276
C2_B_b_NTP	277
C1_B_b_NTP	277
C2_Dec	277
Chry_d12	280
C1_B_b_ThPh	280
C3_Dec	282
C3_B_b_NTP	283
C4_Chry	287
C2_B_b_ThPh	287
cis_Dec	287
C4_Dec	288
C4_DBT	289
B_b_ThPh	291
B_a_Fluorant	293

FIGURE 13.4   Features which has more than 80% of missing value (Program result).

In the current experiment with the three imputers, the RMSE values for the algorithms as

1. Mean: 26.659 (Figure 13.5)
2. Median: 24.558 (Figure 13.6)
3. KNN: 14.694 (Figure 13.7)

```
          OBJECTID  Naph_d8  Naph_d8_imputed
161          235     83.97         76.708634
162          237     65.39         76.708634
163          238     53.79         76.708634
164          239     85.23         76.708634
165          241     65.96         76.708634
166          242     74.34         76.708634
167          243     87.10         76.708634
168          245     59.32         76.708634
169          246     59.14         76.708634
170          248     74.45         76.708634
171          249     69.25         76.708634
172          250     96.85         76.708634
173          252     96.22         76.708634
174          253     86.54         76.708634
175          254    102.95         76.708634
176          255     86.96         76.708634
177          256     89.01         76.708634
178          257     89.46         76.708634
179          258     89.40         76.708634
180          259     85.25         76.708634
181          260     98.54         76.708634
182          261     85.76         76.708634
183          262     83.64         76.708634
184          263     84.00         76.708634
185          264     58.13         76.708634
186          265     79.76         76.708634
187          266     81.56         76.708634
188          267     81.61         76.708634
189          268     87.43         76.708634
190          269     80.01         76.708634
OBJECTID                253.400000
Naph_d8                  80.700667
Naph_d8_imputed          76.708634
dtype: float64
OBJECTID                254.500000
Naph_d8                  83.985000
Naph_d8_imputed          76.708634
dtype: float64
Root Mean Square Error:  26.659114351358763
```

FIGURE 13.5   Imputation using mean (Program result).

From the above results, we can see the KNN has the least RMSE value for the imputation of the "Naph_d8" feature. So, we mark KNN as the imputation technique for this column. Similarly, we can extend this approach to each and every column to be considered in our feature engineering. This iterative approach can be used to identify the most suitable algorithm for the imputation of that feature.

Now, for outliers management, as from Figure 13.8 and Figure 13.9, it can be seen that the undesired outlier of "OBJECTID-222" of "Moisture" is replaced by the maximum value of 89.22 using the proposed method discussed earlier in this chapter.

	OBJECTID	Naph_d8	Naph_d8_imputed
161	235	83.97	79.1
162	237	65.39	79.1
163	238	53.79	79.1
164	239	85.23	79.1
165	241	65.96	79.1
166	242	74.34	79.1
167	243	87.10	79.1
168	245	59.32	79.1
169	246	59.14	79.1
170	248	74.45	79.1
171	249	69.25	79.1
172	250	96.85	79.1
173	252	96.22	79.1
174	253	86.54	79.1
175	254	102.95	79.1
176	255	86.96	79.1
177	256	89.01	79.1
178	257	89.46	79.1
179	258	89.40	79.1
180	259	85.25	79.1
181	260	98.54	79.1
182	261	85.76	79.1
183	262	83.64	79.1
184	263	84.00	79.1
185	264	58.13	79.1
186	265	79.76	79.1
187	266	81.56	79.1
188	267	81.61	79.1
189	268	87.43	79.1
190	269	80.01	79.1

```
OBJECTID             253.400000
Naph_d8               80.700667
Naph_d8_imputed       79.100000
dtype: float64
OBJECTID             254.500
Naph_d8               83.985
Naph_d8_imputed       79.100
dtype: float64
Root Mean Square Error:  24.558455497382205
```

FIGURE 13.6   Imputation using median (Program result).

```
       OBJECTID  Naph_d8  Naph_d8_imputed
104       235     83.97            79.544
105       237     65.39            68.902
106       238     53.79            68.166
107       239     85.23            82.996
108       241     65.96            71.194
109       242     74.34            77.406
110       243     87.10            72.900
111       245     59.32            66.698
112       246     59.14            66.698
113       248     74.45            81.806
114       249     69.25            76.772
115       250     96.85            69.860
116       252     96.22            88.530
117       253     86.54            86.840
118       254    102.95            89.682
119       255     86.96            83.652
120       256     89.01            89.764
121       257     89.46            87.814
122       258     89.40            86.698
123       259     85.25            79.544
124       260     98.54            90.724
125       261     85.76            87.964
126       262     83.64            84.468
127       263     84.00            81.248
128       264     58.13            69.860
129       265     79.76            80.932
130       266     81.56            82.802
131       267     81.61            80.390
132       268     87.43            82.640
133       269     80.01            82.802
OBJECTID               253.400000
Naph_d8                 80.700667
Naph_d8_imputed         79.976533
dtype: float64
OBJECTID               254.500
Naph_d8                 83.985
Naph_d8_imputed         81.527
dtype: float64
Root Mean Square Error:  14.694161671641785
```

FIGURE 13.7 Imputation using KNN (Program result).

#	A	B	C	D	E	F	G	H	I	J	K	L	M	N	O	P	Q	R	S	T	U	V	W
205	1.3E+07	3897167	204	L-S14-12:Sediment	SDC01-ED	Sediment (San Diego	Cardiff Sta	33.01403	-117.281	2012/08/2	17:25:00	2012/08/3	2012/12/1	2013/01/25	00:00:00	1000		8.1414	18.7	0.6141	ng/g (ppxi)	Dry Weight	
206	1.3E+07	3897142	205	L-S14-12:Water	SDC1-ED	Surface W San Diego	San Diego U	33.01385	-117.28	2012/08/2	17:44:00	2012/08/2	2012/12/1	2013/01/25	00:00:00			8.0516	21.7	0.621	ng/g	Dry Weight	0.005 μg/L (ppxi)
207	1.3E+07	3897142	206	L-S14-12:Sediment	SDC1-ED	Sediment (San Diego	San Diego U	33.01385	-117.28	2012/08/2	17:51:00	2012/08/2	2012/12/1	2013/01/25	00:00:00	1000		7.9279	21.1	0.6307	ng/g	Dry Weight	
208	1.3E+07	3897142	207	L-S14-12:Water	SDC1-ED	Surface W San Diego	San Diego U	33.01385	-117.28	2012/08/2	17:56:00	2012/08/2	2012/12/1	2013/01/25	00:00:00								0.005 μg/L (ppxi)
209	1.3E+07	3863385	208	L-S14-12:Sediment	SDF01-ED	Sediment (San Diego	Mission Be	32.7592	-117.254	2012/08/3	12:44:00	2012/08/3	2012/08/3	2012/09/1		1000							
210	1.3E+07	3363385	209	L-S14-12:Sediment	SDF01-ED	Sediment (San Diego	Mission Be	32.7592	-117.254	2012/08/3	12:50:00	2012/08/3	2012/12/1	2013/01/25	00:00:00			9.7838	7.7	0.511	ng/g (ppxi)	Dry Weight	
211	1.3E+07	3363385	210	L-S14-12:Sediment	SDF01-ED	Sediment (San Diego	Mission Be	32.7592	-117.254	2012/08/3	12:55:00	2012/08/3	2012/12/1	2013/01/25	00:00:00			8.1612	21.7	0.6127	ng/g (ppxi)	Dry Weight	
212	1.3E+07	3363385	211	L-S14-12:Tissue	SDF01-ED	Mussel Tis San Diego	Mission Be	32.7592	-117.254	2012/08/3	12:42:00	2012/08/3	2012/10/0	2012/11/15	00:00:00			1.3616	86.8	3.6722	ng/g (ppxi)	Dry Weight	5.89
213	1.4E+07	4979671	212	L-S19-13:Water	HMGS012I	Surface W Humboldt	Eureka Ter	40.77813	-124.196	2013/11/1	14:16:00	2013/11/1	2013/11/1	2013/12/1		1000				0.02-0.05	μg/L (ppxi)		0.00854
214	1.4E+07	4968660	213	L-S19-13:Tissue	HMGS006I	Oyster Tis Humboldt	Coast Seal	40.83925	-124.126	2013/11/1	08:00:00	2013/11/1	2014/06/0	2014/07/2		10.25	1.99875	80.5	4494	ng/g(ppxi)	Dry Weight	2.151345	
215	1.4E+07	4984128	214	L-S19-13:Tissue	HMGS011I	Mussel Tis Humboldt	Samoa Bri	40.80844	-124.155	2013/11/1	11:44:00	2013/11/1	2014/06/0	2014/07/2		10.47	1.4658	86	13.6-34.1	ng/g(ppxi)	Dry Weight	4.03875	1
216	1.4E+07	4976941	215	L-S19-13:Tissue	HMGS001I	Mussel Tis Humboldt	Humboldt	40.75956	-124.222	2013/11/1	14:57:00	2013/11/1	2014/06/0	2014/07/2		10.491	1.961817	81.3	10.2-25.5	ng/g(ppxi)	Dry Weight	2.915664	
217	1.4E+07	4979681	216	L-S19-13:Sediment	HMGS012I	Sediment (Humboldt	Eureka Ter	40.7782	-124.194	2013/11/1	14:37:00	2013/11/1	2014/05/0	2014/06/1		10.346	8.6079	16.8	2.3-5.8	ng/g(ppxi)	3.567655	3.770967	1
218	1.4E+07	4979669	217	L-S19-13:Sediment	HMGS012I	Sediment (Humboldt	Eureka Te	40.77812	-124.194	2013/11/1	14:47:00	2013/11/1	2014/05/0	2014/06/1		10.453	9.710837	7.1	2.1-5.1	ng/g(ppxi)	2.485883	2.825709	1
219	1.4E+07	4977757	218	L-S19-13:Sediment	HMGS001I	Sediment (Humboldt	Coast Gua	40.76518	-124.221	2013/11/1	13:48:00	2013/11/1	2014/05/0	2014/06/1		10.067	9.231439	8.3	2.2-5.4	ng/g(ppxi)	2.028936	5.524599	2
220	1.4E+07	4977767	219	L-S19-13:Sediment	HMGS001I	Sediment (Humboldt	Coast Gua	40.76518	-124.221	2013/11/1	13:54:00	2013/11/1	2014/05/0	2014/06/1		10.288	9.814752	4.6	2.0-5.1	ng/g(ppxi)	2.337298	6.786609	3
221	1.4E+07	4991210	220	L-S19-13:Sediment	HMGS007I	Sediment (Humboldt	Arcata Ma	40.85658	-124.098	2013/11/1	15:59:00	2013/11/1	2014/05/0	2014/06/1		10.232	5.504816	46.2	3.6-9.1	ng/g(ppxi)	18.76357	55.58406	3
222	1.4E+07	4991276	221	L-S19-13:Sediment	HMGS007I	Sediment (Humboldt	Arcata Ma	40.85703	-124.098	2013/11/1	16:21:00	2013/11/1	2014/05/0	2014/06/1		10.257	5.3644	47.7	3.7-9.3	ng/g(ppxi)	20.98832	58.87878	3
223	1.4E+07	4976647	222	L-S19-13:Sediment	HMGS013I	Sediment (Humboldt	Stinky Bea	40.7562	-124.197	2013/11/1	14:35:00	2013/11/1	2014/05/0	2014/06/1		10.452	10.452	**89.22**	1.9-4.8	ng/g(ppxi)	Dry Weight	2.867394	1
224	1.4E+07	4976647	223	L-S19-13:Sediment	HMGS013I	Sediment (Humboldt	Stinky Bea	40.7562	-124.197	2013/11/1	14:45:00	2013/11/1	2014/05/0	2014/06/1		10.408	8.8052	15.4	2.3-5.7	ng/g(ppxi)	Dry Weight	2.687058	1
225	1.4E+07	4976642	224	L-S19-13:Sediment	HMGS013I	Sediment (Humboldt	Elk River B	40.75616	-124.195	2013/11/1	15:23:00	2013/11/1	2014/05/0	2014/06/1		10.315	8.6337	16.3	2.3-5.8	ng/g(ppxi)	Dry Weight	0.606927	
226	1.4E+07	4991165	225	L-S19-13:Water	HMGS007I	Surface W Humboldt	Arcata Ma	40.85627	-124.099	2013/11/1	09:32:00	2013/11/1	2013/12/1	2013/12/1		1000				0.02-0.05	μg/L (ppxi)		0.00759
227	1.4E+07	4976642	226	L-S19-13:Sediment	HMGS013I	Sediment (Humboldt	Elk River B	40.75616	-124.196	2013/11/1	15:25:00	2013/11/1	2014/05/0	2014/06/1		10.018	8.7858	12.3	2.3-5.7	ng/g(ppxi)	Dry Weight	0.883245	0
228	1.4E+07	4979671	227	L-S19-13:Tissue	HMGS012I	Mussel Tis Humboldt	Eureka Ter	40.77813	-124.22	2013/11/1	14:10:00	2013/11/1	2014/06/0	2014/07/2		10.058	2.001542	80.1	4494	ng/g(ppxi)	Dry Weight	3.702146	
229	1.4E+07	4977745	228	L-S19-13:Water	HMGS001I	Surface W Humboldt	Coast Gua	40.76503	-124.22	2013/11/1	14:15:00	2013/11/1	2013/11/1	2013/12/1		1000				0.02-0.05	μg/L (ppxi)		0.01334
230	1.4E+07	4981520	229	L-S19-13:Water	HMGS012I	Surface W Humboldt	Del Norte	40.79071	-124.188	2013/11/1	11:59:00	2013/11/1	2013/11/1	2013/12/1		1000				0.02-0.05	μg/L (ppxi)		0.02454
	⋯	⋯	⋯														⋯		⋯			⋯	⋯

FIGURE 13.8 Replacement of outlier value (Program result).

222	OBJECT ID
L-619-13-25	Lab_Num
Sediment	Sample_Typ
HMGS013ED2111313S D1a	Sample_ID
Sediment Core(12"	Medium
Humboldt Bay	**Region**
StinkyBeach(upper interdital)	Location
40.7562	Longitude
-124.1974	Longitude
2013/11/13	Date_Col
14:35:00	Time_Col
2013/11/15	Date_Rec
2014/05/10	Date_Ectrc
2014/06/10	Date_Anayl
10.452	Mass_Vol_E
10.452	Dry_Samp_ Mass
89.22	**Moisture**
1.9-4.8	RL
Nglg(ppb) Dry Weight	Measure
NaN	Naph
2.867394	F2_MeNaph

FIGURE 13.9   New row of *OBJECTID-222* (Program result).

13.7   CONCLUSION AND FUTURE WORK

Data cleaning is essential to make analysis accurate and for error-free machine learning models; otherwise, whole efforts will be ruined. For intelligent data analysis, data quality has a major role. Since the datasets of standard data sources are already clean, it becomes a little tough to apply the proposed work. In this chapter, "Naph_d8" and "Moisture" features are chosen for missing data handling and outlier management operations respectively. As most of the features in this dataset are chemical compounds, for the same operations to be done with other feature columns, one needs to find correlation coefficients first. Depending on the coefficient, appropriate methods and algorithms have to be used. In this experiment, 300 real data points with 118 features were used. As real data increases, the data training will be more powerful, which will be the future scope of this chapter. So far, single-source data is considered, whereas multi-source data handling on the same project is challenging.

However, a new data cleansing technique for single-source and multi-source-based data quality problems is described by Erharo Harm et al. in [12]. It may also be further applied to get more structured and unique data.

REFERENCES

[1]   Mawdsley, J. et al. 2011. *Renewable vs. Hydrocarbon:The Energy Reality*. Institutional Research AltraCrop Inc.

[2]   Mironov, Oleg G. 1968. Hydrocarbon pollution of the sea and its influence on marine organisms. *Institute of Biology of the South Seas, A.S. Ukr. SSR. SSR Sevastopol, USSR. Helgol~inder wiss. Meeresunters 17,* 335–339.

[3]   Floodgate, G. D. 1995. Some environmental aspects of marine hydrocarbon bacteriology. *School of Ocean Sciences. Marine Science Laboratories, University of Wales 9,* 3–11.

[4]   Seifert, J.W. 2004. Resources science and Industry division. *CRC report for congress. Order code. R131978.1–11.*

[5]   Warner, T.A. 2004. Geobotanical and Lineament analysis of Landsat Satellite Imagery for Hydrocarbon Microseep. Information Bridge, DOE scientific and technical information.

[6]   Ridzuan, F. et al. 2019. A Review on Data Cleansing Methods for Big Data. *Procedia Computer Science 116,* 731–738.

[7]   Devi, Sapna et al. 2015. Study of data cleaning and comparison of data cleaning tools. *IJCSMC 4*(3), 360–370.

[8]   Lakshni, S. et al. 2018. An overview study on data cleaning, its type and its methods for data mining. *International Journal of Pure and Applied Mathematics 119*(12), 16837–16848.

[9]   Manimekalai, K. et al. 2018. Missing value imputation and normalization techniques in myocardial infarction. *ICTACT Journal of Soft Computing 8*(3).

[10] Song, Qinbao. et al. 2007. Missing data imputation techniques. *International Journal of Business Intelligence and Data Mining 2*(3), 261–291.

[11] Wangikar, Vaishali Chandrakant. et al. 2011. Data cleaning: Current approaches and issues. *IEEE International Conference on Knowledge Engineering.*

[12] Rahm, Erharo *et al.* Data cleaning: Problems and Current Approaches. *IEEE Data Engineering Bulletin.*

Index

Note: Figures are indicated by *italics*. Tables are indicated by bold.

Printed in the United States
by Baker & Taylor Publisher Services